BEATING OXIDATIVE STRESS NATURALLY

Dr. Scott A. Johnson

Beating Oxidative Stress Naturally / Scott A. Johnson

Cover design: Scott A. Johnson
Cover Copyright © 2026, by Scott A. Johnson

ISBN-13: 979-8988720683

Published by Scott A. Johnson Professional Writing Services, LLC: Orem, UT

Discover more books by Scott A. Johnson at authorscott.com/shop/

Contents

CHAPTER ONE

Unraveling Oxidative Stress

Modern life places extraordinary demands on the human body. Every breath we take, every calorie we burn, and every stress we endure leaves behind invisible byproducts that quietly shape how we feel, age, and function. At the center of this process is oxidative stress—a natural phenomenon that becomes harmful when the body is pushed beyond its ability to repair and protect itself. While oxidative stress is not a disease in itself, it acts like a slow-burning fuse, influencing everything from energy levels and brain function to inflammation, immune resilience, and the pace at which our bodies wear down over time. Many of today's most common health struggles—fatigue, brain fog, chronic pain, accelerated aging, and degenerative disease—share a common thread that often goes unrecognized: the cumulative damage caused by unchecked oxidative stress.

What makes oxidative stress especially important—and empowering—is that it is not merely something that "happens" to us. It is shaped every day by the foods we eat, the air we breathe, the technology we use, the sleep we get, and even the way we think and respond to stress. Your body is equipped with sophisticated defense systems designed to neutralize damage and restore balance, but these systems require the right support to function optimally. This book exists to bridge the gap between complex science and practical living, helping you understand why oxidative stress matters and, more importantly, what you can do about it. By learning how to restore balance naturally—through nutrition, lifestyle choices, targeted support, and mindful living—you gain the tools to protect your health at the cellular level and reclaim a sense of vitality that many assume is lost with age. This is not about fear—it's about awareness, control, and the opportunity to change the trajectory of your health starting now.

Oxidative stress is a biological process that occurs when there is an imbalance between the production of free radicals and the body's ability to neutralize them with antioxidants. Free radicals are unstable molecules that contain unpaired electrons, making them highly reactive. To stabilize themselves, they steal electrons from other molecules, including proteins, lipids, and DNA. The stealing of electrons from nearby molecules leads to a chain reaction that can result in cellular damage. While free radicals are naturally produced in the body as part of normal cellular functions, excessive levels can contribute to aging, inflammation, and a range of diseases, including cancer, neurodegenerative disorders, and cardiovascular disease.

Redox balance is a term that refers to the balance between two opposing forces in the body: oxidation and reduction. Oxidation is a process where a molecule loses electrons, and during this process, it often produces free radicals. Imagine oxidation as a tiny fire that can spread and cause chaos in a crowded room if not contained. On the other hand, reduction involves gaining electrons, which can help to stabilize molecules. This is where antioxidants come in. Antioxidants are substances that can donate electrons to free radicals, helping to neutralize them and prevent them from causing damage—like putting out the tiny fire before it turns into a larger blaze. The best way to imagine this may be to think in terms of a kitchen fire. If your pot of dinner catches fire on the stove, putting it out quickly prevents the fire from spreading to the adjacent kitchen cabinets and walls and avoids a major house fire.

Now, imagine that this kitchen fire isn't a one-time accident, but something that happens repeatedly throughout the day. A small flare-up here, a spark there—on the stove, in the oven, or near the toaster. In a well-run kitchen, there are fire extinguishers, smoke alarms, and someone paying attention, ready to act the moment a flame appears. These safeguards represent your body's antioxidant systems. But if fires start faster than they can be put out, or if the extinguishers are empty, broken, or missing, damage quickly spreads. Cabinets scorch, walls blacken, and eventually the entire kitchen becomes unsafe. This is oxidative stress: a state where the body's natural fire-control systems are overwhelmed, allowing small, routine sparks to turn into widespread damage that affects how the entire "house" functions. Over time, that accumulated damage begins to show up as fatigue, inflammation, accelerated aging, and disease—clear signs that the fires have been burning unchecked for far too long.

So, in essence, redox balance is all about maintaining harmony between these two processes. When everything is balanced, your cells function optimally, and your body stays healthy. But if oxidative stress sets in, it can lead to damage and is linked to a variety of health issues, such as aging, heart disease, and diabetes. Therefore, keeping the redox balance in check—by consuming a healthy diet rich in antioxidants, managing stress, staying active, and using targeted supplementation—is crucial for maintaining overall health and well-being.

Picture atoms and molecules as tiny building blocks. They prefer to exist with their electrons (even smaller particles) paired up. When a molecule loses an electron, it becomes unstable and reactive, meaning it wants to steal an electron from something else to become stable again. That's a free radical. This electron stealing can damage other molecules.

There are two main categories of free radicals, including:

Reactive Oxygen Species (ROS)

The most common type of free radical your body encounters are ROS. They're like oxygen molecules that have gone slightly wild. We all need oxygen to breathe, and it is a calm and beneficial molecule. However, sometimes it can become excited or unstable and turn into ROS. Think of it like oxygen that has been given a jolt of energy by drinking an energy drink. This makes it restless and unstable, and because its unstable, it begins relentlessly pursuing a missing piece of itself (a lost electron). Once it encounters a molecule that has an electron, it steals it, which damages the other molecule. There are many forms of ROS the body has to deal with:

- *Superoxide.* This is one of the first free radicals formed during normal cell processes. It's like a slightly agitated oxygen molecule.
- *Hydroxyl Radical.* This is a very aggressive free radical. It's a tiny, extremely angry molecule that can cause a lot of damage.
- *Hydrogen Peroxide.* You might know this as a disinfectant. While not technically a free radical itself, it can turn into those more dangerous hydroxyl radicals.
- *Peroxyl Radicals.* These are formed when fats in your body react with oxygen. They're important in the process of lipid peroxidation, which is basically when fats go rancid inside your body.
- *Singlet Oxygen.* This is a form of oxygen that's been energized. It's especially reactive to things like lipids and DNA.

ROS can also influence the activity of matrix metalloproteinases (MMPs). MMPs are a group of enzymes in the body that help break down proteins in the extracellular matrix—the network of proteins and other substances that provide structure and support to tissues. MMPs play a vital role in many processes, including wound healing, tissue remodeling, and the normal development of organs. However, these enzymes need to be tightly regulated because if they are overactive, they can break down too much tissue, leading to problems like arthritis (breakdown cartilage in joints), heart disease (remodel blood vessels contributing to atherosclerosis and aneurysms), impaired wound healing (breakdown too much extracellular matrix, hindering proper tissue formation), and even certain cancers (promote tumor invasion and metastasis).

Visualize your body's tissues as needing regular maintenance, and MMPs act like tiny, helpful tools that perform this maintenance. These tools are essential for keeping everything in good shape and are usually well-regulated and. Now, imagine tiny reactive substances called ROS. A small amount of these substances is actually necessary for the tools to work correctly. However, when there's an overabundance of ROS, it's like throwing fuel onto a fire, causing the MMPs to become overactive. This can happen in a few ways: the excess ROS can signal the body to produce more MMPs, it can energize the existing MMPs making them work more aggressively, or it can interfere with the body's natural way of controlling the MMPs, which involves substances called tissue inhibitors of metalloproteinases (TIMPs). Normally, TIMPs act like supervisors, ensuring the MMPs don't overdo it. But too much ROS can weaken these supervisors. Consequently, while MMPs are crucial for maintaining healthy tissues under normal conditions, an excess of ROS can lead to uncontrolled MMP activity, resulting in unwanted damage and inflammation within the body's tissues. In essence, a delicate balance is needed: MMPs are vital, but their activity can spiral out of control in the presence of too many reactive oxygen species, ultimately contributing to various health problems.

Reactive nitrogen species (RNS) are highly reactive molecules that contain nitrogen and can cause damage to cells and tissues in the body. They are produced naturally during various biological processes, especially when the immune system is fighting off infections or inflammation. While they play important roles in signaling and helping the body to respond to stress or pathogens, excessive amounts of RNS can lead to harmful effects, similar to reactive oxygen species (ROS). This can result in damage to DNA, proteins, and lipids, which may contribute to various diseases, including cardiovascular problems and neurodegenerative conditions. Balancing these reactive molecules is crucial for maintaining good health, as both too much and too little activity can be problematic. Two forms of RNS exist:

- *Nitric Oxide.* This one's tricky because it requires a delicate balance. In small amounts, it's actually helpful, but too much can be harmful. On the positive side, it helps regulate blood flow by relaxing blood vessels, supports the immune system by fighting off infections, and acts as a signaling molecule in various physiological processes. However, when produced in excessive amounts or inappropriately, it can lead to harmful effects. High levels of nitric oxide can react with other molecules, forming damaging compounds that contribute to inflammation, tissue injury, and various diseases.

- *Peroxynitrite.* This is formed when nitric oxide reacts with superoxide, and it's a very damaging free radical. Peroxynitrite is also like a double-edged sword. While it can play some roles in signaling and defense in small amounts, it is often considered harmful because it is very reactive and can damage important cellular components like proteins, lipids, and DNA. When peroxynitrite levels become too high, it can contribute to various health problems, including inflammation and diseases like arthritis, heart disease, and neurodegenerative conditions.

Free radicals are produced in the body during normal metabolic processes, particularly in the mitochondria, the energy-producing structures within cells. As mitochondria generate ATP (the energy currency of the cell) through a process called oxidative phosphorylation, electrons are transferred along the electron transport chain. The electron transport chain is like a tiny power plant inside your cells. It takes high-energy electrons from food molecules and passes them down a line of protein complexes, like a bucket brigade. As these electrons move, they release energy, which the power plant uses to pump protons (tiny, charged particles) across a membrane, creating a concentration difference, like water built up behind a dam. Finally, these protons flow back across the membrane through a special enzyme, ATP synthase, which acts like a turbine, using the flow to generate ATP, the cell's main energy currency. Oxygen is the final receiver of the low-energy electrons at the end of the chain, combining with them and protons to form water. Occasionally, some of these electrons escape and react with oxygen to form reactive oxygen species (ROS), such as superoxide and hydrogen peroxide. While these molecules have beneficial roles in cell signaling and immune defense, excessive production or inadequate antioxidant protection can lead to oxidative stress. The normal cellular processes that produce free radicals are:

- *Metabolism.* When our cells convert nutrients from food into energy, they generate free radicals as byproducts. This is most prominent in the mitochondria, the energy-producing structures within cells.

- *Immune Responses.* When the body detects pathogens or harmful agents, immune cells produce free radicals to help destroy these invaders. This is a crucial part of the body's defense mechanism, but an overactive immune response can contribute to oxidative stress. Free radicals are also used to defend against cancer. Immune cells, like tiny soldiers, intentionally produce a burst of these reactive substances when they encounter cancerous cells. This sudden surge of free radicals and oxidative stress damages the cancerous cells, destroying them or hindering their growth.

- *Inflammatory Processes.* During inflammation, immune cells are activated and can produce high levels of free radicals. While this is a necessary part of healing, chronic inflammation can lead to sustained oxidative stress.

In addition to the free radicals our bodies naturally produce, we are also exposed to free radicals through various environmental and dietary sources. Air pollution, cigarette smoke, radiation, industrial chemicals, and heavy metals can all contribute to oxidative stress by introducing reactive molecules into the body. Diet also plays a major role, as the consumption of processed foods, excessive sugar, unhealthy fats, and alcohol can increase oxidative burden. Cooking methods such as frying and grilling at high temperatures generate advanced glycation end products (AGEs) and lipid peroxides, which contribute to oxidative stress when consumed in large amounts. On the other hand, a diet rich in antioxidants—such as those found in colorful fruits and vegetables, nuts, seeds, and herbs—helps counteract oxidative damage and supports overall health. Lack of physical activity and poor quality or too little sleep can also lead to higher levels of oxidative stress.

When oxidative stress occurs, either by too many free radicals being produced or having too few antioxidants to neutralize them, it can wreak havoc on cellular structures, leading to significant biological damage through several mechanisms. Understanding these mechanisms—lipid peroxidation, DNA damage, and protein modification—sheds light on how oxidative stress contributes to many diseases and aging.

Lipid peroxidation is one of the most damaging processes initiated by free radicals. Lipids in cell membranes, predominantly polyunsaturated fatty acids, are particularly susceptible to oxidative damage. When free radicals attack these lipids, they steal electrons, resulting in the formation of lipid radicals. These radicals can further react with oxygen to produce peroxyl radicals, which can propagate a chain reaction, continuously attacking nearby lipids.

This cascading effect leads to the formation of lipid peroxides, which can disrupt the integrity of cell membranes by increasing permeability and fluidity. The consequences of lipid peroxidation are profound, as compromised membranes can lead to cellular dysfunction. For example, the loss of membrane integrity can result in the leakage of intracellular contents and the entry of harmful substances, ultimately leading to cell death. Cell death occurs when a cell can no longer perform its essential functions or maintain its internal structure, essentially shutting down and being removed by the body. While some cell death is normal, excessive or premature cell death can impair tissue function, weaken organs, and contribute to inflammation, aging, and the development of chronic disease.

Moreover, lipid peroxidation generates secondary products, such as malondialdehyde (MDA) and 4-hydroxynonenal (4-HNE), which are highly reactive. These byproducts can further modify proteins and DNA, creating a vicious cycle that exacerbates oxidative damage. Thus, lipid peroxidation not only disrupts cellular structure but also initiates pathways leading to inflammation, tissue damage, and even premature apoptosis (programmed cell death).

Another critical mechanism of damage from oxidative stress is DNA damage. Free radicals can directly attack the nucleobases in DNA, leading to a variety of injuries, including single-strand breaks, double-strand breaks, and modifications of the bases themselves. A well-studied modification is the formation of 8-oxoguanine, derived from the oxidation of guanine (G).[1,2] This modified base can result in mispairing during DNA replication, leading to mutations that can accumulate over time. DNA is a precise instruction manual for your cells, written with four letters (bases), including guanine. Sometimes, damage from oxidation can change a G into an "8-oxoG," which is like a typo in the manual. When the cell tries to copy the DNA with this typo, it can misread the 8-oxoG and put in the wrong letter, leading to a mutation. Over time, these typos can build up.

DNA damage can trigger a cascade of cellular events, including cell cycle arrest, apoptosis, or senescence (where cells stop dividing). Cell cycle arrest is like hitting the pause button—when a cell detects damage, it temporarily stops dividing so it can attempt repairs before continuing. If the damage is too severe to fix, the cell may undergo apoptosis, which is a form of programmed cell death where the cell safely shuts itself down and is removed to protect surrounding tissue. Senescence occurs when a damaged cell doesn't die but permanently stops dividing and remains in the body, often releasing inflammatory signals that can disrupt nearby cells. While these processes are protective in the short term, excessive DNA damage can overwhelm these systems, contributing to tissue dysfunction, aging, and disease over time. The accumulation of DNA mutations is a hallmark of aging[3] and is implicated in various diseases, including cancer.[4,5] Importantly, the body is equipped with built-in repair systems—such as base excision repair and nucleotide excision repair pathways—designed to identify and fix DNA damage before it becomes permanent. However, when oxidative damage is excessive or ongoing, these repair systems can become overwhelmed, allowing mutations to accumulate and leading to genomic instability—meaning the DNA becomes increasingly error-prone, fragile, and unreliable, making cells more likely to malfunction or turn cancerous over time.

Base excision repair and nucleotide excision repair are two important processes that cells use to fix damaged DNA, which can occur due to environmental factors or normal cellular activities. In base excision repair, the cell identifies and removes specific damaged bases (the building blocks of DNA), like those altered by oxidation or chemicals. Once the faulty base is removed, an enzyme fills in the gap by adding the correct base, and a DNA polymerase seals the area to restore the DNA strand. This is somewhat like a sprinkler pipe repair. If your DNA is a long sprinkler pipe carrying genetic information (water), sometimes, a section of the pipe gets damaged, or a wrong part gets installed (a damaged base). Base excision repair is like a skilled plumber that spots the leak or incorrect fitting in the pipe, cuts it out carefully, installs the correct pipe and fittings, then seals the section so the flow of information (water) continues smoothly without any leaks or errors.

On the other hand, nucleotide excision repair is used for fixing larger sections of DNA that contain more extensive damage, like bulky adducts or distortions caused by UV light. In this process, the cell cuts out a short segment of DNA that includes the damaged nucleotides, and then new nucleotides are added to fill the gap, followed by sealing the strand. Both repair pathways are crucial for maintaining the integrity of our DNA and preventing mutations that could lead to diseases like cancer. We can connect this process to the modern activity of road repair. Your DNA acts as a long, smooth highway. Nucleotide excision repair is like a road crew that identifies a section of the highway with multiple potholes or a significant crack (representing a bulky DNA lesion). Instead of just patching each pothole individually, the crew comes in and removes the entire damaged segment of the road. Then, they lay down a brand new, correctly paved section, ensuring a smooth and continuous flow of traffic (the cell's processes) once again. This repair mechanism deals with larger, more disruptive damage that affects multiple "bases" (road units) at once.

To understand these repair processes better, think of your DNA as a long, intricate necklace made of beautiful beads, where each bead represents a DNA base. Over time, this necklace might get damaged; some beads might become discolored or broken due to exposure to sunlight or getting snagged on something sharp.

Base excision repair is like a careful jeweler who inspects the necklace and notices a single discolored bead. Instead of replacing the entire necklace, the jeweler gently removes just that damaged bead and replaces it with a new, perfectly matching one, making the necklace as good as new.

On the other hand, nucleotide excision repair is akin to a situation where a section of the necklace has multiple broken beads and perhaps some missing pieces due to an accident. In this case, the jeweler decides to cut out a larger segment containing the damaged beads. Then, he replaces that entire section with fresh, vibrant beads and secures everything back together, restoring the necklace to its original beauty.

Both repairs are vital for ensuring that the necklace remains intact and attractive, just as these DNA repair processes are crucial for keeping our genetic information safe and functional.

The implications of DNA damage are particularly relevant in the context of cancer. Oxidative stress-induced mutations can activate oncogenes—genes responsible for cell growth. When these genes become stuck in the "on" position or become overactive due to mutations, they can trigger cancer cells to grow and divide uncontrollably. These mutations can also inactivate tumor suppressor genes, which contributes to the initiation and progression of tumors.[6] Consequently, chronic exposure to oxidative stress is recognized as a contributor to carcinogenesis, illustrating the importance of understanding how oxidative stress impacts DNA integrity.

Proteins, essential for nearly all cellular functions, are also vulnerable to oxidative damage. Free radicals can modify amino acid residues, particularly those containing sulfur, such as cysteine and methionine. This process can lead to the formation of disulfide bonds, protein unfolding, or aggregation, impairing the protein's functional capacity. Disulfide bonds form when sulfur-containing parts of a protein become chemically "stuck" together in the wrong places, altering the protein's shape. Protein unfolding occurs when a protein loses its precise three-dimensional structure—much like a carefully folded piece of origami being crumpled—preventing it from doing its job. Aggregation happens when damaged proteins begin clumping together instead of staying separate and functional, creating bulky masses that interfere with normal cellular operations.

One notable example of protein modification due to oxidative stress is carbonylation, which can occur when free radicals attack amino acids, resulting in the formation of carbonyl groups. Carbonylated proteins tend to lose their activity and can become targets for proteasomal degradation. In simple terms, carbonylation acts like a "damage tag" placed on a protein, signaling that it is no longer functioning properly. The proteasome functions as the cell's cleanup and recycling system—a molecular shredder that recognizes these damaged proteins, breaks them down, and allows their components to be reused. The accumulation of damaged proteins can disrupt numerous cellular processes, including signaling pathways, metabolic functions, and the structural integrity of the cytoskeleton (the internal framework of a cell, made of a network of protein fibers).

The link between protein modification and disease is well-established. In neurodegenerative diseases, such as Alzheimer's and Parkinson's, oxidative modification of proteins contributes to the pathology observed in these conditions.[7] Proteins can misfold or aggregate, leading to the formation of toxic species that interfere with neuronal function. Therefore, the oxidative modification of proteins is a critical mechanism by which oxidative stress can lead to cellular dysfunction and disease. In reality, these misfolded proteins may occur due to dysfunction in brain glucose utilization and insulin sensitivity.[8,9] In both diseases, specific proteins misfold and clump together, like amyloid-beta and tau in Alzheimer's, and alpha-synuclein in Parkinson's. This buildup seems to disrupt the normal way brain cells, especially neurons and astrocytes, process glucose. Think of it like a traffic jam preventing fuel from reaching the brain cells that need it. The brain can become less responsive to insulin, a condition called insulin resistance, which impairs glucose entrance into cells to be used for energy. If the brain becomes resistant to insulin, it can't take up glucose as efficiently, leading to an energy shortage. This energy deficit can further impair the proper folding of proteins, creating a vicious cycle where misfolding worsens and energy production suffers. So, protein clumping and problems with glucose use and insulin sensitivity appear to be intertwined in the progression of these neurodegenerative diseases.

To envision this process of protein modification, imagine a complex, intricate clockwork mechanism (representing a protein) essential for the smooth operation of a factory (the cell). This clockwork is made of many precisely shaped gears and levers (amino acids). Free radicals are like corrosive rust or grit. They attack the delicate parts of the clockwork, particularly those made of softer metals (sulfur-containing amino acids). This rust/grit can cause the gears to stick together (disulfide bond formation), bend out of shape (protein misfolding), or clump together (protein aggregation), making the clockwork malfunction. Free radicals leave behind these "sticky residues" (carbonyl groups), further hindering the clockwork's movement and marking it for removal. If enough clockwork mechanisms are damaged, the entire factory's production slows down, becomes erratic, or even grinds to a halt (disruption of cellular processes). In a factory specializing in precise, delicate work (neurons), damaged clockwork leads to the production of faulty parts (toxic protein species), ultimately causing the factory to break down (neurodegeneration). In essence, just as rust and damage disrupt a complex machine, oxidative stress disrupts proteins, leading to cellular dysfunction and disease.

Understanding the distinction between acute and chronic oxidative stress is essential for appreciating how oxidative stress affects health over time. Both forms can have significant biological impacts, but their nature, duration, and underlying mechanisms differ.

Acute oxidative stress refers to a short-term and often reversible cellular response to intense stressors such as infection, inflammation, or environmental toxins. This type of oxidative stress typically occurs during acute physiological events—like a vigorous exercise session or an immune response to pathogens—where there is a transient increase in ROS production. While excessive ROS can cause cellular damage in the short term, acute oxidative stress can also trigger adaptive responses.

For instance, during moderate exercise, the body temporarily experiences increased oxidative stress due to heightened metabolic activity. However, this acute increase in ROS can stimulate the expression of endogenous antioxidants and enhance the body's overall antioxidant defenses, contributing to improvements in fitness and health. In this sense, acute oxidative stress can activate beneficial adaptations, but the balance between oxidative damage and repair capacity is crucial. If the oxidative stress is too intense or prolonged, it can lead to tissue injury and associated pathologies.

Conversely, chronic oxidative stress is characterized by a long-term imbalance between the production of ROS and the body's ability to neutralize them through antioxidants. It typically arises from persistent factors such as poor diet, environmental pollutants, sedentary lifestyle, chronic inflammation, and age-related decline in antioxidant defenses. Unlike acute oxidative stress, which is often transient, chronic oxidative stress can lead to ongoing cellular damage and contribute to the development of various diseases.

Chronic oxidative stress is implicated in a plethora of pathological processes. For example, oxidative stress has been recognized as a critical player in the pathogenesis of chronic inflammatory diseases, cardiovascular diseases, neurodegenerative disorders, and cancer.[10,11,12] The prolonged exposure to free radicals leads to cumulative damage—resulting in mitochondrial dysfunction, impaired DNA repair systems, and inflammation—all of which can create a vicious cycle of oxidative stress and tissue injury.

Moreover, chronic oxidative stress can affect the signaling pathways involved in cellular homeostasis, resulting in altered cell proliferation, survival, and apoptosis. Over time, the cumulative effects of oxidative damage contribute to aging and various age-related diseases, emphasizing the critical nature of managing oxidative stress over a lifetime.

Oxidative stress's mechanisms of damage—including lipid peroxidation, DNA damage, and protein modification—highlight the profound implications of free radical activity on cellular health. Differentiating between acute and chronic oxidative stress further underscores the complexity of oxidative damage, illustrating how acute responses may lead to beneficial adaptations, while chronic imbalance can contribute to a spectrum of diseases. Understanding these processes emphasizes the importance of maintaining antioxidant defenses through lifestyle choices, thereby promoting long-term health and mitigating the effects of oxidative stress.

The body has evolved sophisticated antioxidant defense systems to counteract oxidative stress and prevent cellular damage. These include enzymatic and non-enzymatic antioxidants. Enzymatic antioxidants include superoxide dismutase (SOD), which converts superoxide radicals into hydrogen peroxide; catalase (CAT), which breaks hydrogen peroxide down into water and oxygen; and glutathione peroxidase (GPX), which neutralizes hydrogen peroxide and lipid peroxides using glutathione, a powerful endogenous antioxidant. Non-enzymatic antioxidants include molecules such as glutathione, uric acid, coenzyme Q10, and melatonin, which directly scavenge free radicals and prevent oxidative damage. These antioxidants work together in a network, recycling each other to maintain an effective and robust defense system.

Superoxide Dismutase (SOD)

Role and Function. SOD is one of the first lines of defense against oxidative stress. It catalyzes the conversion of superoxide radicals ($O_2 \cdot -$) into hydrogen peroxide (H_2O_2) and molecular oxygen (O_2). Superoxide is a byproduct of various metabolic processes, and its accumulation can be harmful. By converting superoxide into less harmful compounds, SOD significantly reduces oxidative damage within cells.

Cofactors. SOD requires metallic cofactors—small mineral helpers like zinc, copper, or manganese—to function effectively. There are different isoforms of SOD, such as Copper/Zinc-SOD found in the cytosol—the fluid component enclosed within the cellular membrane, and Manganese-SOD located within mitochondria. Copper (Cu), zinc (Zn), and manganese (Mn) are the essential metal ions that act as cofactors for their respective SOD isoforms, assisting in electron transfer during the reaction.

Catalase (CAT)

Role and Function. Catalase is an enzyme that further processes hydrogen peroxide, a product of the SOD reaction. It rapidly decomposes H_2O_2 into water (H_2O) and oxygen (O_2), preventing the accumulation of hydrogen peroxide, which can be toxic at high concentrations. Catalase plays a critical role in protecting cells from oxidative damage and maintaining redox balance.

Cofactors. Catalase predominantly relies on iron (Fe) as a cofactor for its enzyme activity. Iron is integral to the heme group within the catalase structure, facilitating the enzymatic breakdown of hydrogen peroxide.

Glutathione Peroxidase (GPX)

Role and Function. GPX is another crucial enzyme in the antioxidant arsenal. It catalyzes the reduction of hydrogen peroxide and organic peroxides to water and alcohol, respectively. GPX utilizes glutathione (GSH), a potent endogenous antioxidant, as a reducing agent in its reactions. By detoxifying harmful peroxides, GPX protects cellular membranes and proteins from oxidative damage.

Cofactors. The enzyme requires selenium (Se) as a cofactor, specifically in the form of selenocysteine. This essential trace element is critical for the structure and function of several GPX isoforms, enabling their antioxidant activity.

Antioxidants do not operate in isolation; rather, they work together in a synergistic network to optimize cellular protection. For instance, glutathione, through its various forms and interactions with enzymes like GPX, helps maintain the redox state of the cell, supporting overall antioxidant capacity.

The body's antioxidant defense systems include crucial enzymatic antioxidants. These enzymes, produced within the body, play a vital role in neutralizing free radicals and reducing oxidative stress. SOD, CAT, and GPX work tirelessly to defend us against free radical assaults. Each of these enzymes require specific cofactors or nutrients to function optimally. By maintaining a well-balanced diet rich in the cofactors and nutrients necessary for these enzymatic antioxidants— such as minerals like zinc, copper, and selenium—we can bolster these defense mechanisms and promote better health and longevity.

In addition to the endogenous antioxidants, your body makes several other non-enzymatic antioxidants—such as uric acid, coenzyme Q10 (CoQ10), and melatonin—that play crucial roles in mitigating oxidative stress within the body.

Each of these molecules possesses unique properties and functions that contribute to overall cellular health.

Glutathione, Reduced (GSH)

Role and Function. Glutathione is a tripeptide composed of three amino acids: cysteine, glutamine, and glycine. It is often referred to as the body's master antioxidant due to its ability to scavenge free radicals, neutralize peroxides, and detoxify harmful substances. Glutathione is also vital for the function of other antioxidants, such as GPX, and plays a key role in the maintenance of the immune system. Glutathione contains sulfur (specifically, thiol groups -SH), which have a strong affinity for certain metals and toxins. This affinity means it is "sticky" and can bind to harmful substances, forming complexes that can then be eliminated from the body. This binding process is a form of chelation, where glutathione essentially grabs onto the metal or toxin, making it crucial for detoxification.

Cofactors. The synthesis of glutathione relies on specific amino acids (cysteine, glutamine, and glycine) and may require adequate levels of selenium for optimal GPX activity, as it further utilizes glutathione for its protective effects.

Uric Acid

Role and Function. Primarily known for its association with gout, uric acid is a naturally occurring waste product generated from the breakdown of purines, which are components found in various foods and in the body's cells. At moderate levels, uric acid acts as an antioxidant, effectively scavenging free radicals and protecting cells from oxidative damage. Uric acid is particularly known for its role in reducing oxidative stress in the vascular system, where it helps to preserve endothelial function and prevent damage related to inflammation.

Cofactors. Uric acid does not rely on specific cofactors for its antioxidant activity. However, its production and metabolism may be influenced by dietary factors, hydration status, and the presence of other antioxidants in the body. High levels of uric acid can lead to inflammation and gout, so balance is key.

Coenzyme Q10 (CoQ10)

Role and Function. Coenzyme Q10, also known as ubiquinone, is a lipid-soluble antioxidant that plays a vital role in the electron transport chain within mitochondria, where it facilitates the production of adenosine triphosphate (ATP), the primary energy currency of cells. It's important to understand that CoQ10 exists in two forms—ubiquinone and ubiquinol. Ubiquinone is the oxidized form produced by the body and ubiquinol is the reduced, active antioxidant form. The body converts ubiquinone to ubiquinol to leverage its antioxidant properties. CoQ10 not only functions as an essential component in energy metabolism but also acts as a powerful antioxidant that protects cellular membranes by neutralizing lipid peroxides and preventing oxidative damage to lipid structures.

In addition to its antioxidant function, CoQ10, in its reduced form (ubiquinol), helps regenerate other antioxidants, such as vitamin E, thereby enhancing the overall antioxidant defense system. Its ability to mitigate oxidative damage is critical for maintaining mitochondrial health and cellular energy levels.

Cofactors. CoQ10 synthesis relies on several nutrients, particularly vitamins B6, B12, folic acid, and the amino acid tyrosine. Deficiencies in these nutrients can impair CoQ10 production, reducing its availability for both energy production and antioxidant protection.

Melatonin

Role and Function. Melatonin is a hormone primarily secreted by the pineal gland in response to darkness. While it is best known for regulating sleep-wake cycles, melatonin also exhibits significant antioxidant properties. It directly scavenges a variety of reactive species, including hydroxyl radicals, hydrogen peroxide, and peroxynitrite, which are potent sources of oxidative damage.

Melatonin enhances the expression of various antioxidant enzymes, such as SOD and GPX, thereby boosting the body's overall antioxidant capacity. Furthermore, melatonin has been found to stabilize cellular membranes and prevent lipid peroxidation, supporting brain health and neuroprotection.

Cofactors. Melatonin synthesis is influenced by several nutrients, including the amino acid tryptophan, which serves as its precursor. This synthesis pathway also requires vitamin B6, magnesium, and zinc. Adequate nutrition and a healthy lifestyle, inclusive of proper sleep hygiene, are essential for optimal melatonin production.

Another mechanism the body uses to combat oxidative stress is metal-binding proteins. Proteins such as ferritin and ceruloplasmin bind to metals that can catalyze free radical formation, thus preventing oxidative damage. Metal binding proteins play an essential role in maintaining cellular homeostasis and protecting against oxidative stress. These proteins not only transport essential metal ions but also help mitigate the potential toxicity of free metals by safely locking them away and regulating their activity. Key metal binding proteins involved in the body's antioxidant defense system include ferritin, transferrin, metallothionein, and ceruloplasmin. Each of these proteins interacts with

21

specific metals, ensuring their proper function while protecting cells from oxidative damage.

Ferritin

Role and Function. Ferritin is a globular protein complex responsible for iron storage and regulation. It consists of a protein shell that can sequester up to 4,500 iron atoms in a non-toxic, soluble form. By keeping extra iron locked away, ferritin helps stop a harmful process called the Fenton reaction. In this process, free iron can mix with a substance called hydrogen peroxide and create dangerous molecules known as hydroxyl radicals. These radicals are highly reactive and can damage our cells. So, by storing excess iron, ferritin prevents this harmful reaction from happening and helps protect our cells from damage. This protective function is critical in preventing oxidative damage to lipids, proteins, and DNA.

Ferritin also plays a role in the release of iron when needed, helping to balance iron availability for essential cellular processes, such as hemoglobin synthesis and energy production.

If you have an autoimmune condition, especially related to the thyroid, you should check iron, ferritin, and transferrin levels when you have bloodwork done. Iron measures the amount of iron currently in your bloodstream. A ferritin blood test shows how much iron is stored in your body. Low levels can indicate insufficient iron stores, while high levels can suggest you have too much iron in your body. Transferrin is a glycoprotein that acts like a delivery truck, carrying iron around in your bloodstream to where it's needed. A transferrin test looks at how well your body can transport iron. If transferrin levels are high, it often means the body is trying to carry more iron because there isn't enough available.

Many autoimmune diseases can affect the body's ability to absorb and use iron properly. Monitoring iron, ferritin, and transferrin helps doctors understand whether someone has sufficient iron levels or if there's an issue causing low iron, leading to symptoms like tiredness or weakness. The thyroid gland helps regulate many body processes, and iron is crucial for the production of thyroid hormones. Low iron can lead to fatigue and problems with thyroid function. Checking all three (iron, ferritin, and transferrin) gives a complete picture of the body's iron status and helps ensure that important bodily functions, including those related to the thyroid and immune system, are running smoothly.

Cofactors. While ferritin itself does not require cofactors to function, its synthesis and regulation depend on adequate levels of iron, as well as involvement from other proteins such as aconitase, which can influence the balance of stored versus free iron in cells. Aconitase is an important enzyme that plays a role in the citric acid cycle (also known as the Krebs cycle), which is a series of chemical reactions used by cells to generate energy from carbohydrates, fats, and proteins. Specifically, aconitase helps convert citrate (a compound derived from carbohydrates) to isocitrate during this energy-producing process.

Aconitase has another important role aside from being an enzyme. It is also a sensor for the levels of iron in the cell. When there is enough iron, aconitase can function as an enzyme in the citric acid cycle. However, when iron levels are low, aconitase can detach from its iron-containing cluster and instead bind to special RNA sequences in the cell that help regulate iron metabolism. When aconitase acts as an iron sensor, it helps control the production of ferritin, the protein that stores iron. If iron levels are high, ferritin production increases to store the excess iron. Conversely, if iron levels are low, aconitase helps reduce ferritin production to make sure iron is available for essential functions in the body.

Transferrin

Role and Function. As mentioned above, transferrin is a glycoprotein responsible for the transport of iron in the bloodstream. Synthesized primarily in the liver, transferrin binds iron ions with high affinity, facilitating their delivery to cells and tissues that require iron, such as the bone marrow for red blood cell production. By binding and transporting iron, transferrin helps reduce the availability of free iron, which can catalyze oxidative reactions leading to cellular damage.

Transferrin also plays a role in regulating inflammation, as it can influence immune responses by modulating the availability of iron to pathogens, thus limiting their growth. When inflammation occurs (like during an infection or an injury), the body often increases the production of a protein called hepcidin, which reduces the amount of iron available in the blood. This is a defense mechanism—by limiting iron, the body makes it harder for bacteria and viruses to thrive as they often need iron to grow. If there's less iron in the bloodstream due to increased hepcidin, transferrin might also decrease because there's less iron to carry. Alternatively, if the body needs to transport more iron to support healing and immune response, transferrin might increase.

Cofactors. Like ferritin, transferrin's functionality is predominantly related to iron. However, the efficiency of transferrin binding can be influenced by the presence of other molecules, including vitamin C, which can enhance iron absorption and bioavailability. Bioavailability describes how much of something you take in actually makes it into your cells, rather than passing through unused—similar to the difference between food you eat and food your body truly absorbs.

Metallothionein

Role and Function. Metallothioneins (MTs) are a family of low-molecular-weight, cysteine-rich proteins that bind heavy metals, such as zinc, copper, and cadmium, as well as protect against oxidative stress. MTs can maintain metal ion homeostasis, facilitate detoxification, and regulate metal availability for biological processes. They exhibit a high capacity for binding to various metal ions, which allows them to mitigate the toxicity of excess metals and contribute to antioxidant defenses by scavenging free radicals.

Moreover, metallothioneins are particularly important during oxidative stress because the body produces more of them when cells sense oxidative damage, increasing their ability to bind excess metals and help restore balance inside the cell.

Cofactors. The synthesis of metallothioneins is influenced by zinc and other metals, while certain hormones—such as glucocorticoids, insulin, estrogen, thyroid hormones—and cytokines can also regulate their expression. Additionally, optimal production may depend on adequate intake of sulfur-containing amino acids, particularly cysteine.

Ceruloplasmin

Role and Function. Ceruloplasmin is a copper-carrying protein synthesized in the liver, functioning primarily as an enzyme with oxidase activity. It binds copper ions (Cu^{2+}), facilitating their transport in the bloodstream while simultaneously preventing copper toxicity. Ceruloplasmin has antioxidant properties as it can oxidize iron from its ferrous (Fe^{2+}) to ferric (Fe^{3+}) form, thereby promoting iron incorporation into transferrin and reducing free iron levels, which helps mitigate oxidative stress. By regulating copper and iron balance, ceruloplasmin plays a crucial role in protecting cells from iron-induced oxidative damage and supporting overall antioxidant defenses.

Cofactors. Ceruloplasmin function is directly related to the availability of copper for its synthesis. Additionally, other antioxidants, such as vitamin C and glutathione, can influence the oxidative state within which ceruloplasmin operates.

Metal binding proteins, including ferritin, transferrin, metallothionein, and ceruloplasmin, are integral components of the body's defense against oxidative stress. Each of these proteins serves distinct but complementary roles in managing metal ion availability and protecting against the harmful effects of free metals and oxidative species. By sequestering metals, transporting essential nutrients, and modulating oxidative processes, these proteins contribute significantly to cellular homeostasis and overall health.

With a better understanding of oxidative stress, the natural question becomes: how do you know if you are experiencing it, or if it is contributing to your symptoms? Several blood tests and biomarkers can indicate oxidative stress in humans. These markers help assess oxidative damage to lipids, proteins, and DNA, as well as the body's antioxidant defense capacity. Here are some of the most commonly used biomarkers:

Markers of Oxidative Damage

Lipid Peroxidation Products

- Malondialdehyde (MDA)—A byproduct of lipid peroxidation, commonly measured via TBARS (thiobarbituric acid reactive substances) assay.
- F2-Isoprostanes—Stable markers of lipid oxidation, considered one of the most reliable indicators of oxidative stress.
- 4-Hydroxynonenal (4-HNE)—A toxic aldehyde formed by lipid peroxidation in cells.

Protein Oxidation Products

- Advanced Oxidation Protein Products (AOPPs)—Reflect oxidative damage to proteins.
- Protein Carbonyls—Indicators of protein oxidation, often increased in oxidative stress-related conditions.

DNA Oxidation Products

- 8-Hydroxy-2'-deoxyguanosine (8-OHdG)—A marker of oxidative DNA damage, often measured in blood or urine.

Markers of Antioxidant Defenses

Enzymatic Antioxidants

- Superoxide Dismutase (SOD)—Converts superoxide radicals into hydrogen peroxide and oxygen.
- Catalase (CAT)—Breaks down hydrogen peroxide into water and oxygen.
- Glutathione Peroxidase (GPX, or GPx)—Reduces hydrogen peroxide and lipid peroxides.

Non-Enzymatic Antioxidants

- Total Antioxidant Capacity (TAC)—Measures the overall antioxidant status in plasma or serum.

- Glutathione Ratio (GSH/GSSG ratio)—The reduced (GSH) to oxidized (GSSG) glutathione ratio is a key indicator of oxidative stress.
- Vitamin C and Vitamin E—Essential antioxidants that can be measured in the blood.

Inflammation and Oxidative-Stress Related Markers

- C-Reactive Protein (CRP)—Elevated levels suggest inflammation, often linked to oxidative stress.
- Myeloperoxidase (MPO)—An enzyme released by white blood cells that contributes to oxidative stress and inflammation.
- Homocysteine—Elevated levels are associated with increased oxidative stress and cardiovascular risk.

Below is a table summarizing the oxidative stress biomarkers along with their normal reference ranges (where available). Keep in mind that normal ranges can vary depending on the laboratory, testing method, and individual factors such as age, gender, and health status.

BIOMARKER	TYPE	NORMAL RANGE	NOTES
Lipid Peroxidation Products			
Malondialdehyde (MDA)	Lipid oxidation	$0.5 - 5\ \mu mol/L$	Measured via TBARS assay
F2-Isoprostanes	Lipid oxidation	$10 - 100\ pg/mL$	Considered a reliable marker of oxidative stress
4-Hydroxynonenal (4-HNE)	Lipid oxidation	$< 1\ \mu mol/L$	Elevated in oxidative damage
Protein Oxidation Products			
Advanced Oxidation Protein Products (AOPPs)	Protein oxidation	$50 - 200$ $\mu mol/L$	Higher levels indicate oxidative stress
Protein Carbonyls	Protein oxidation	$0.1 - 0.5$ nmol/mg protein	Associated with aging and diseases

BIOMARKER	TYPE	NORMAL RANGE	NOTES
DNA Oxidation Products			
8-Hydroxy-2'-deoxyguanosine (8-OHdG)	DNA oxidation	0.2 – 5 ng/mL (serum/plasma) or <10 ng/mg creatinine (urine)	Common marker of DNA oxidative damage
Enzymatic Antioxidants			
Superoxide Dismutase (SOD)	Antioxidant enzyme	1.5 – 3.5 U/mL	Converts superoxide radicals into oxygen and hydrogen peroxide
Catalase (CAT)	Antioxidant enzyme	20 – 500 kU/L	Decomposes hydrogen peroxide
Glutathione Peroxidase (GPx)	Antioxidant enzyme	6 – 20 U/g Hb	Reduces peroxides and protects cells
Non-Enzymatic Antioxidants			
Total Antioxidant Capacity (TAC)	Antioxidant status	1.0 – 2.5 mmol Trolox equivalents/L	Measures the combined antioxidant activity of all antioxidants
Glutathione (GSH/GSSG ratio)	Antioxidant status	>10:1 (GSH:GSSG)	A lower ratio indicates oxidative stress
Vitamin C (Ascorbic Acid)	Antioxidant vitamin	34 – 114 µmol/L	Water-soluble antioxidant
Vitamin E (α-Tocopherol)	Antioxidant vitamin	12 – 42 µmol/L	Fat-soluble antioxidant

BIOMARKER	TYPE	NORMAL RANGE	NOTES
Inflammation and Oxidative Stress-Related Markers			
C-Reactive Protein (CRP)	Inflammation	< 1 mg/L (low risk)	Elevated in oxidative stress and inflammation
Myeloperoxidase (MPO)	Inflammation/oxidative stress	10 – 200 pM	Involved in oxidative damage in the cardiovascular system
Homocysteine	Oxidative stress and cardiovascular risk	5 – 15 µmol/L	Elevated levels linked to oxidative stress and cardiovascular disease

Ultimately, oxidative stress is a natural part of life, but maintaining a balance between free radical production and antioxidant defense is key to promoting long-term health and reducing the risk of chronic disease. Understanding how to support the body's natural defense systems through diet, lifestyle, supplements, essential oils, and environmental choices can help mitigate the negative effects of oxidative stress and promote longevity.

CHAPTER TWO

EMF: A Hidden Danger Amplifying Oxidative Stress

In recent years, concern about electromagnetic frequencies (EMFs) and health has increased significantly. As technology becomes an ever more integral part of our daily lives, EMFs have become ubiquitous, emanating from devices such as smartphones, Wi-Fi routers, laptops, microwaves, and even power lines. These electromagnetic fields are a natural part of our environment; however, the proliferation of electronic devices has resulted in our exposure to higher levels of EMF than ever before.

Electromagnetic fields are invisible areas of energy, often called radiation, that surround us. They are produced by electrically charged objects and consist of both electric fields and magnetic fields, hence the name electromagnetic. They are generated by both natural and manmade sources. Natural sources include the Earth's magnetic field and atmospheric static electricity. Manmade sources of EMFs include power lines, electrical appliances (microwaves, computers, televisions, etc.), cell phones, wireless devices, radios, and broadcast equipment.

EMFs are categorized by their frequency. Non-ionizing radiation has lower frequencies and doesn't have enough energy to break chemical bonds in atoms. Examples include radio waves, microwaves, and the fields from power lines. Ionizing radiation involves higher frequencies, like X-rays and gamma rays, and can damage cells by breaking chemical bonds. While non-ionizing EMFs are generally regarded as safe at established exposure levels, as the presence of such fields increases, so do public concerns about potential long-term health effects.

As research continues to evolve, some studies suggest that prolonged exposure to EMFs might be linked to a range of health issues, including sleep disturbances, headaches, fatigue, and more serious conditions such as cancer.[13] These concerns are compounded by the rapid rollout of advanced technologies, including 5G networks, which promises faster internet speeds and enhanced connectivity but raises new questions about the safety of higher-frequency EMFs. As communities embrace these technologies, debates regarding the potential risks and benefits are becoming more pronounced, prompting further investigation into how EMFs interact with biological systems.

Moreover, the widespread adoption of personal devices has led to significant changes in lifestyle, with individuals often spending extended periods in close proximity to EMF sources. Many people now carry smartphones in their pockets and use wireless devices throughout the day, making it nearly impossible to avoid exposure entirely. This reality has encouraged a growing movement advocating precautionary measures, such as reducing EMF exposure when possible and promoting research that seeks to understand the implications of long-term exposure on human health. The discussion surrounding EMFs encompasses not only scientific considerations but also societal and regulatory aspects, as policymakers grapple with how to balance technological advancements with public health concerns.

The potential association between electromagnetic fields (EMF)—such as those from Wi-Fi, 5G, and other wireless technologies—and oxidative stress has been a topic of research and debate in recent years. While some studies suggest a link, the overall evidence remains inconclusive, and more research is needed to establish clear connections.

Some studies in cell cultures and animal models have indicated that exposure to EMF can lead to an increase in ROS, disrupting the body's natural balance and leading to oxidative stress.[14,15,16,17] These studies often report changes in antioxidant levels, DNA damage, and cellular function after continuous or high levels of EMF exposure. The emerging evidence suggests that EMFs may overwhelm our antioxidant defenses, leading to oxidative stress and increasing the risk of oxidative stress-related health conditions.

Studies on animals, mainly rats and mice, indicate that electromagnetic field (EMF) exposure leads to oxidative stress in multiple organs.[18] The brain, testes, liver, and kidneys are the most commonly affected organs. Many studies focus on the effects of EMFs on the brain since mobile phones are held close to the head during use.

A review article concluded that EMFs from everyday technology can have negative effects, particularly on the nervous system.[19] The available evidence pointed to the reality that EMFs may interfere with brain function and potentially contribute to neurological issues. This can partly be explained by their effects on the heat shock response. Our cells produce special stress proteins (heat shock proteins or HSPs) when exposed to harmful environmental factors like heavy metals, UV radiation, or EMFs. HSPs are like the body's emergency response team when cells are under stress. These proteins help the body adapt to stress but can also signal cell damage.

When EMFs expose the brain to oxidative stress, increased temperature, or disruptions in cellular function, the body activates HSPs to help cells survive. However, this stress response may also indicate that cells are struggling to cope with damage. HSPs are normally protective, but if they're constantly activated due to chronic EMF exposure, it means cells are repeatedly stressed. Over time, this can weaken cells, making them more vulnerable to damage and dysfunction. EMFs trigger free radical production, which can damage brain cells (neurons). If oxidative stress overwhelms the brain, neurons can die off, leading to issues with memory, focus, and cognitive function.

Additionally, EMFs can disrupt the brain's protective barrier, allowing toxins and heavy metals to enter brain tissue.[20] This can lead to inflammation, oxidative stress, and even neurodegenerative diseases. With a weakened blood-brain barrier, metals like iron, aluminum, and mercury can accumulate in the brain, contributing to oxidative stress and neurotoxicity. When unwanted molecules pass into the brain, the immune system responds with inflammation, causing further damage to brain cells.

The brain is especially sensitive to oxidative stress, due to limited renewal and short lifespan. Some research links EMF exposure to headaches, memory issues, sleep disturbances, and even neurodegenerative diseases like Alzheimer's.[21,22] A review article noted that increased ROS production was observed in rodent brains after exposure to radiofrequency (RF) and extremely low-frequency (ELF) EMFs.[23] This oxidative stress was linked to impaired learning, memory deficits, and signs of neurodegeneration, especially in older animals. This makes sense since older mammals tend to have diminished antioxidant systems capacity and activity. Emerging evidence suggests that EMF exposure, even in lower doses, may alter cellular oxidative balance, triggering adaptive responses after recovery, and, in some cases, pathological effects on nervous system function.

Other preclinical data suggests that EMF exposure interferes with the electrical balance in cells.[24] The primary findings were that EMFs disrupt the normal movement of charged atoms (cations) across cell membranes and affect the function of channels that control electrical signals. This disruption can cause an increase in oxidative stress, which can damage various cell functions, including DNA, and potentially lead to cancer. The study specifically examined how EMFs influence an enzyme called NADPH oxidase (NOX) to produce oxidative stress.

Essentially, the EMFs disrupt the electrical signals cells use to communicate with each other, triggering more ROS to be made, and an oxidative stress response that harms cells.

EMFs have the potential to interfere with reproductive functions as well. This is exactly what a review article found in preclinical models of both male and female reproductive systems.[25] Research shows that EMF exposure triggers the disproportionate production of ROS by the mitochondria of reproductive cells—excessive leakage from the mitochondrial electron transport chain—leading to reproductive harm. Within the mitochondria, there's a process called the electron transport chain. It's a series of steps that produce energy. However, this process isn't perfect and can sometimes leak harmful byproducts (ROS). In small amounts, ROS are normal, but when there's excessive leakage from the electron transport chain, too many ROS are produced. This can create oxidative stress, which can damage highly sensitive sperm in men—particularly the Y chromosome—and egg cells in women. In men, excessive ROS can damage sperm, affecting the sperm's ability to swim, their DNA, and their overall quality. In women, oxidative stress can harm egg cells, potentially affecting their development and ability to be fertilized. So, when there's too much leakage in the mitochondria's energy production process, it can lead to a buildup of harmful molecules that damage the delicate reproductive cells in both men and women, potentially causing problems with fertility.

The effects of EMF have also been evaluated in rats on their antioxidant system in the liver, heart, kidneys, and plasma. Exposure to extremely low frequency electromagnetic fields for two hours—notably less than the average human is exposed to these frequencies during an average day—reduced GSH activity in the heart and kidney and SOD in plasma.[26] Another study concluded that chronic EMF exposure is harmful to the kidneys.[27] Higher creatinine and urea nitrogen levels in the blood were observed, indicating that the kidneys aren't filtering waste as efficiently as they should. These substances are normally removed from the blood by healthy kidneys, so an increase means the kidneys might be struggling. Kidney blood flow was also restricted or backed up after EMF exposure, signifying inflammation or damage. Lastly, a reduction in red blood cell count in glomerular capillaries—a unique network of small blood vessels located within the glomerulus of the kidney—was observed, possibly suggesting insufficient red blood cells were reaching the kidney's tiny blood vessels, which could decrease overall kidney function. In a nutshell, the study findings suggest that EMFs could damage the kidneys leading to toxin buildup in the blood (which could cause fatigue, nausea, or serious kidney issues), elevated blood pressure, and increased risk of kidney disease.

Another study compared the effects of ionizing and non-ionizing radiation on the liver of rats.[28] Three of four liver enzymes (AST, ALT, ALP) were elevated after exposure to both types of EMF, suggesting liver function was negatively impacted. The researchers hypothesized that oxidative stress and inflammation-related pathways were involved in the liver function alterations.

Human clinical research is currently sparse and has produced mixed results. For example, the INTERPHONE international case-control study attempted to assess brain tumor risk associated with mobile phone use by reviewing 2,708 glioma and 2,409 meningioma cases and matched controls from thirteen countries.[29] Overall, no increased risk of the two brain tumors was observed with mobile phone use. There were potential indications of increased glioma risk at the highest exposure levels, but study biases and errors prevented causal interpretation. Therefore, this study was inconclusive and recommended further investigation.

Other research reviews have looked at a broader connection between mobile phone use and various types of cancer. Analysis of nine epidemiological studies available at the time from five different countries, estimated a relative risk of increased cancers between 1.6 and 4.6 times.[30] The highest overall risk was for acoustic neuroma (3.5 times) and uveal melanoma (4.2 times). A more recent review and meta-analysis found evidence that linked mobile phone use—cumulative call time of more than 1,000 hours—to increased tumor risk.[31] The research findings seem to be trending towards a link between cancer and EMF exposure from mobile phones.

A meta-analysis published in 2014 evaluated if an association existed between magnetic field exposure and childhood leukemia.[32] A total of 11,699 childhood leukemia cases and 13,194 controls from nine studies were stratified by different exposure cut-off points. Based on the analysis, a statistical association between magnetic field exposure and childhood leukemia was observed—a 31% increased risk of developing leukemia. A more recent study determined that the actual odds for getting childhood B-lineage acute lymphoblastic leukemia increased 6% when living in a city with elevated exposure to extremely low-frequency magnetic fields.[33] Finally, a review and meta-analysis of 38 studies investigating the link between EMF and childhood leukemia determined exposure to extremely low-frequency magnetic fields is associated with a 27% pooled increased risk of developing leukemia in childhood, most commonly acute lymphoblastic leukemia.[34] Based on the current evidence, an association between childhood leukemia and chronic EMF exposure is pretty clear.

Sleep is incredibly important for overall health. It plays a vital role in processing and consolidating memories. Lack of sleep impairs the brain's ability to transfer information from short-term to long-term memory, which can lead to difficulties with learning, concentration, and recall. The brain is cleansed during sleep as well, with the space between neurons opening up slightly to allow fluid to clear away debris, plaques, and misfolded proteins. Sleep helps regulate emotions, so sleeping poorly can increase irritability, mood swings, and vulnerability to stress, anxiety, and depression. During sleep, the immune system produces cytokines, which help fight infection and inflammation. Sleep deficiency weakens the immune system, making individuals more susceptible to illness. Sleep is essential for regulating hormones that control appetite, metabolism, and growth. Lack of sleep can disrupt these hormonal balances, increasing the risk of obesity, diabetes, and other metabolic disorders. Sleep helps regulate blood pressure and heart rate. Chronic sleep deprivation increases the risk of heart disease, high blood pressure, and stroke. Lastly, the body repairs tissues and releases growth hormones during sleep, and insufficient sleep can hinder these processes. It is abundantly clear that sleep quantity—7–9 hours for adults per night—and quality—time spent in sleep stages: N1(light sleep) 5%; N2 (deeper light sleep) 45%–55%; N3 (deep restorative sleep) 15%–25%; and REM sleep 20%–25% —is critical to optimum health.

One cross-sectional study evaluated the sleep quality of people who worked in an electrical plant. Interestingly, the daily occupational exposure to EMF was associated with reduced sleep quality, increasing poor sleep quality risk by 83%–112%.[35] Another study reported similar findings in power plant workers, with the worst sleep quality experienced among those with the highest EMF exposure.[36] Additionally, they noted higher depression rates among the EMF group, with direct effects on stress, anxiety, and depression. Contrarily, another study found that sleep quality was not affected by mobile phone use or daily radiofrequency electromagnetic field exposure.[37] And another study determined that whole-night Wi-Fi exposure by sleeping next to a router did not reduce sleep quality, despite slight changes in alpha frequency brainwaves caused by Wi-Fi.[38] The EMF connection to sleep quality is inconclusive and warrants additional research.

Many subjects report nervous system symptoms such as headaches, fatigue, and cognitive difficulties when around EMF sources. Others report skin problems, such as rashes or facial irritation, as well as nonspecific health-related symptoms. Dubbed electromagnetic hypersensitivity (EHS), but also known as microwave syndrome, three explanatory hypotheses have been proposed.[39] The first is the electromagnetic hypothesis, which directly attributes sensitivity and symptoms to EMF exposure. Second, some assume the sensitivity and symptoms are simply a nocebo effect based on false beliefs that EMF is harmful. Lastly, one hypothesis suggests that symptoms and sensitivity are a coping strategy for other preexisting conditions.

There is a strong connection between oxidative stress and EHS.[40] In fact, studies show that approximately 80 percent of people with EHS have one or more measurable oxidative stress biomarkers in their blood.[41] This means their symptoms are not subjective or imagined—they are accompanied by objective biological signs that the body is under oxidative strain.

EHS also overlaps significantly with another condition known as multiple chemical sensitivity (MCS). People with MCS experience adverse reactions to very low levels of everyday chemicals that most individuals tolerate without issue. These triggers may include fragrances, cleaning products, pesticides, smoke, fumes, and certain building materials. About 30 percent of EHS cases are linked to MCS, and in over one-third of these combined cases, MCS appears first and is later followed by EHS. This overlap suggests that the two conditions may share common underlying biological mechanisms rather than being separate or unrelated disorders.

Researchers believe these shared mechanisms involve disruptions in how the nervous system, immune system, and detoxification pathways function. One possible factor is central nervous system sensitization, where the brain and nerves become overly reactive and begin responding too strongly to stimuli that would normally be harmless. Another is autonomic nervous system dysregulation, meaning the system that controls automatic functions—such as heart rate, digestion, and stress responses—becomes unbalanced, often leaving the body stuck in a chronic "fight-or-flight" state. Chronic inflammation may also play a role, in which the immune system remains activated for too long and continues damaging tissues instead of resolving the original trigger. Additionally, impaired detoxification can make it harder for the body to efficiently process and eliminate toxins, allowing environmental chemicals to linger longer and provoke symptoms.

The immune system appears to be involved as well. Nearly one-quarter of individuals with EHS and MCS have been found to carry autoantibodies,[42] which are immune proteins that mistakenly attack the body's own tissues rather than foreign invaders. In this case, these autoantibodies target myelin, the protective coating around nerves. Myelin functions much like insulation around electrical wires—when it is damaged, nerve signals can become distorted or disrupted, potentially contributing to heightened sensitivity, pain, and neurological symptoms.

To better understand what is happening in the brain, researchers have also examined blood flow and brain activity in people with EHS using advanced imaging techniques. They discovered impaired circulation in a major brain artery called the middle cerebral artery, along with reduced activity in a region known as the capsulo-thalamic area of the temporal lobes. This region is closely connected to the limbic system, which governs emotions and memory, and the thalamus, which plays a central role in sensory processing and alertness. In addition, markers of inflammation and signs of a compromised blood-brain barrier—the protective filter that normally shields the brain from toxins—were observed.

Taken together, these findings paint a consistent picture. The combination of oxidative stress, immune dysfunction, nervous system sensitization, impaired brain blood flow, and inflammation strongly suggests that EHS is a neurological condition with measurable biological underpinnings. Recognizing EHS as such is essential—not only for proper diagnosis and treatment, but also for prevention strategies aimed at reducing oxidative burden and protecting the nervous system before long-term damage occurs.

An often-overlooked source of EMF is electric vehicles. Just like your phone, a microwave, or even the power lines outside your house, anything that uses electricity creates EMFs. The high-voltage battery packs, electric motors and power inverters, regenerative braking systems, wiring and battery managements systems, and particularly the charging systems all contribute to EMF exposure.

Scientists and health groups have set safety guidelines for EMFs, and electric cars are designed to meet these limits. Investigations have shown that the levels of EMFs in electric cars are often similar to or even lower than what you'd experience from many household items. However, it's worth noting that some experts believe these safety limits might not be strict enough for very long-term exposure—like daily trips in your electric vehicle—especially because a part of the World Health Organization has suggested that very low-frequency magnetic fields could possibly be linked to cancer.[43] Moreover, over 250 scientists have signed the International EMF Scientist Appeal, calling for stronger guidelines.[44] While electric cars do produce EMFs, and some scientists are calling for more research into long-term effects, the general consensus from current studies is that the levels are within safe guidelines for regular use. Still, while many studies find no conclusive evidence linking low-level, non-ionizing EMFs from EVs to adverse health effects, the cumulative effect of long-term exposure to various low-level EMFs remains a subject of ongoing inquiry and debate.

The research on EMF exposure and health is diverse and ongoing, and while some studies suggest potential health effects, the overall body of evidence remains mixed and inconclusive. Many authorities advocate for continued research, particularly as technology evolves. Individuals concerned about EMF exposure are encouraged to stay informed through credible sources and evaluate their usage patterns where applicable. While a few studies have reported associations between EMF exposure and increased oxidative stress markers, others have found no significant effects. The human body is complex, and individual responses may vary based on factors such as genetic predisposition, duration of exposure, and existing health conditions.

For individuals concerned about EMF exposure and oxidative stress, certain practical steps may help alleviate worries:

- *Monitor Your Surroundings*. Use an EMF meter to assess the levels of exposure in your environment and identify sources of higher EMF.

- *Distance and Duration*. Limit the time spent close to devices emitting EMF when possible. For instance, avoid keeping mobile phones in pockets or near the bed while sleeping. Maintain a safe distance from high-EMF sources like cell phone towers, power lines, and Wi-Fi routers, when possible.

- *Use Wired Connections*. Opt for wired internet connections when feasible instead of Wi-Fi, especially for activities that require prolonged exposure.

- *Use Speakerphone or Wired Headphones*. When making calls, use speakerphone or wired headphones to keep the phone away from your head and body.

- *Text Instead of Calling*. Using text messaging instead of calls can reduce exposure to EMF radiation, particularly in sensitive areas like the brain.

- *Turn Off Devices*. Switch off Wi-Fi, Bluetooth, and mobile data when not in use, especially during sleep hours.

- *Place Wi-Fi Routers Strategically*. Position Wi-Fi routers away from common areas where people spend a lot of time, such as bedrooms and living rooms.

- *Limit Wireless Devices*. Reduce the number of wireless devices in your home. Replace wireless devices with wired ones (e.g., wired printers, desktops).

- *Sleep-EMF Hygiene*. Turn off or unplug devices with sleeping.

- *Use EMF Shielding Products*. Consider using EMF shielding materials or products like bedsheets or canopies specifically designed to reduce EMF exposure during sleep.

- *Avoid Areas with Heavy Wi-Fi Use.* Be cautious in places like coffee shops, libraries, and public transportation where wireless devices and networks are prevalent.
- *Promote Overall Health.* Maintain a healthy lifestyle through diet, exercise, and stress management, which can contribute to your overall well-being and may help mitigate some effects associated with EMF exposure.

In summary, while there is existing and emerging evidence suggesting that EMF exposure may contribute to oxidative stress, more research is needed to fully understand the implications and mechanisms involved. As we navigate a world increasingly integrated with technology, it becomes essential to use this technology wisely to mitigate potential risks while still reaping its numerous benefits.

Adopting a balanced approach encourages us to leverage advancements such as wireless communication, smart home devices, and convenience gadgets without exposing ourselves unnecessarily to higher levels of EMF. For instance, prioritizing the use of wired connections over wireless when appropriate can significantly reduce EMF exposure while maintaining connectivity. Similarly, incorporating regular breaks from screen time and implementing the "distance principle" by keeping devices away from our bodies can further safeguard health.

Moreover, fostering a culture of awareness about EMF sources—whether in the workplace, school, or even at home—can empower individuals to make informed decisions on technology usage. This includes utilizing health-conscious devices that are designed with EMF shielding features, promoting practices such as switching off Wi-Fi during the night, and limiting the wireless devices in one's environment.

Ultimately, as research continues to evolve and provide more clarity on the effects of EMF exposure, it is our responsibility as users to make proactive choices. By being informed about both the risks and advantages of our technological tools, we can create environments that prioritize our health while still embracing the conveniences that modern technology offers. Staying vigilant and adopting safe practices will enable us to navigate our digital landscape more mindfully, fostering well-being in an increasingly electrified world.

CHAPTER THREE

The Harmful Health Effects of Oxidative Stress

Prolonged oxidative stress is killer on the body—literally. It plays a significant role in various health conditions by damaging cells and tissues and disrupting normal cellular and biological functions. Oxidative stress is implicated in aging because accumulated damage to cellular components over time can impair tissue function and contribute to the decline of physiological processes. Many chronic diseases, including Alzheimer's disease, Parkinson's disease, diabetes, and cardiovascular conditions, are linked to excessive oxidative damage as well. The body does have repair mechanisms, such as DNA repair enzymes and autophagy (a process where damaged cellular components are degraded and recycled), but persistent oxidative stress can overwhelm these systems. In case you haven't noticed yet, balance is a cornerstone of good health!

Aging

Oxidative stress is a concept that has gained increasing attention in recent years for its critical role in the aging process and its contribution to various age-related diseases. Over time, the cumulative damage caused by oxidative stress can significantly impact our health, leading to premature aging and even premature death.

To understand how oxidative stress contributes to aging, it is essential to know how our bodies produce energy. Every day, our cells generate energy from the food we eat through a process called cellular respiration. During this process, oxygen is utilized, and in turn, free radicals are produced as byproducts. In small amounts, free radicals play a role in cell signaling and the immune response. However, when they accumulate due to various factors like pollution, poor diet, stress, or a sedentary lifestyle, they can overwhelm the body's antioxidant defenses.

As we age, our ability to combat oxidative stress declines. Several factors contribute to this decline. First, the production of endogenous antioxidants—substances produced by the body to fight free radicals, such as GSH, SOD, and CAT—typically decreases with age.[45] Second, our cells' ability to regenerate and repair themselves diminishes, making it more challenging to recover from oxidative damage.[46] This combination creates a perfect storm where oxidative stress can thrive, accelerating the aging process and making age-related decline more pronounced, widespread, and difficult for the body to recover from over time.

The effects of oxidative stress on the body can be profound. It accelerates the aging process by causing damage to DNA, leading to mutations that disrupt normal cellular functions. This DNA damage can result in the depletion of functional cells, increasing the risk of various age-related diseases. For instance, when cells in critical organs such as the heart, brain, and kidneys are damaged, they can no longer perform their essential functions, leading to conditions like heart disease, neurodegenerative diseases (such as Alzheimer's and Parkinson's), and chronic kidney disease.

Oxidative stress also impacts the body's inflammatory response, which becomes dysregulated with age.[47] Chronic inflammation, often dubbed inflammaging, is characterized by persistent low-grade inflammation that can result from the continuous assault of oxidative stress on tissues. This chronic inflammatory state contributes to the development of age-related diseases, including autoimmune disorders, diabetes, and cardiovascular diseases. Given that these conditions are leading causes of mortality, the role of oxidative stress in promoting them highlights its significance in aging and premature death.

In addition to these cellular and physiological consequences, oxidative stress can also impact our overall health, lifestyle, and mental well-being. For example, oxidative stress is known to affect skin health by promoting premature aging signs like wrinkles and loss of elasticity. The skin is often one of the first places we notice signs of aging, and oxidative damage significantly impacts its appearance and functionality. Beyond appearance, oxidative stress can lead to cognitive decline, impairing memory and cognitive functions, which are vital for maintaining an independent lifestyle as we age.

Notably, oxidative stress can create a cycle of deterioration. As our bodies age and accumulate oxidative damage, we become more susceptible to chronic diseases. These conditions often lead to additional lifestyle limitations, such as reduced physical activity and poor dietary choices, further exacerbating oxidative stress. The stress we encounter in our daily lives, whether from work, relationships, or health concerns, also contributes to the production of free radicals, creating a vicious cycle where stress and aging feed into one another. It's a devastating cycle that severely limits quality of life.

In essence, while chronological age is a fixed measure, biological age and perceived age are influenced by the cumulative effects of oxidative stress and how well an individual's body can counteract it. Those with lower levels of oxidative stress and more effective defense mechanisms may appear younger and healthier than their biological age, while those who struggle to combat oxidative stress look older than they really are.

Oxidative stress is a significant contributor to the aging process and plays a critical role in developing age-related diseases that can lead to premature mortality. The excess of free radicals generated in our bodies over time, particularly as we age, can result in cumulative cellular damage that disrupts normal functioning and drives inflammation. Ultimately, the cumulative damage caused by oxidative stress decreases our overall quality of life and healthspan as we grow older, negatively impacting longevity.

Cancer

As has been discussed, oxidative stress is strongly associated with cancer. Oxidative stress significantly contributes to the development of cancer by causing damage to DNA, proteins, and lipids within cells. The overproduction of ROS can lead to mutations in critical genes that regulate cell growth and division. When the cellular repair mechanisms fail to correct these mutations, it can result in uncontrolled cell proliferation and tumor formation. Additionally, oxidative stress can promote chronic inflammation, which creates an environment conducive to cancer development. Various inflammatory cytokines and growth factors are released in response to oxidative damage, further enhancing the risk of malignancy. Over time, the combination of genetic mutations and a supportive inflammatory environment can lead to cancer progression and metastasis.

Building on the established link between oxidative stress and cancer, it's important to understand that this process isn't just a random occurrence; it's often driven by various external and internal factors. For example, exposure to environmental carcinogens like tobacco smoke and pollutants significantly increases ROS production. The chemicals in tobacco smoke can trigger a cascade of oxidative reactions in lung tissue, increasing the risk of lung cancer. Knowing this, a glaring question arises: why do some smokers never get cancer? The answer to this question is complex but may be due to them possessing more robust DNA repair mechanisms, protective gene variations that help detoxify harmful substances, more effective immune function, enhanced metabolism, or greater antioxidant capacity.

Internal factors like chronic infections and metabolic disorders can also contribute to oxidative stress and cancer development. Chronic infections, such as hepatitis B or C, can cause persistent inflammation and ROS production in the liver, increasing the risk of liver cancer.[48] Similarly, metabolic disorders like diabetes, characterized by high blood sugar levels, can lead to increased ROS production and oxidative damage to various tissues, elevating the risk of cancers like colorectal and pancreatic cancer.[49,50] These internal factors create a sustained environment of oxidative damage, making cells more vulnerable to mutations and uncontrolled growth.

Beyond DNA damage, oxidative stress can also disrupt cellular signaling pathways that are crucial for normal cell function. ROS can modify proteins involved in cell signaling, altering their activity and leading to aberrant cell growth. For instance, the activation of certain signaling pathways, like the NF-κB pathway, by ROS can promote cell survival and proliferation, even in the presence of DNA damage. This pathway is a master regulator of inflammation and cell survival, and when persistently activated, it can drive tumor progression. Research demonstrates that ROS-mediated NF-κB activation plays a critical role in the development of various cancers.[51] Moreover, oxidative stress can also damage lipids, particularly those in cell membranes, leading to the formation of lipid peroxidation products. These products can further contribute to inflammation and DNA damage, creating a vicious cycle that promotes cancer development. Essentially, the multifaceted impact of oxidative stress on cellular components and signaling pathways creates a perfect storm for cancer initiation and progression.

Inflammatory Bowel Disease (IBD)

Inflammatory bowel disease (IBD) is a term that encompasses several inflammatory disorders of the digestive tract, with Crohn's disease and ulcerative colitis being the most common types. These conditions can severely impact the quality of life, causing symptoms such as abdominal pain, diarrhea, and weight loss. One of the key players in the development and progression of IBD is oxidative stress, particularly in the gut.

In a healthy intestine, the epithelial lining acts as a protective barrier, preventing harmful substances, bacteria, and toxins from entering the bloodstream. However, when oxidative stress kicks in, the excessive production of ROS can inflict damage on this delicate intestinal lining. Consider your gut as a well-maintained wall; too much oxidative damage is like constant wear and tear, leading to cracks and weaknesses. This damage disrupts the epithelial barrier, making it more permeable. When the barrier is compromised, harmful substances that are normally kept at bay can enter the bloodstream, triggering an inflammatory response.

This inflammatory response is characterized by the activation of the immune system, which sends in specialized cells to fight off what it perceives as invading threats. However, in the case of IBD, these immune responses can become misguided and chronic. The immune cells release pro-inflammatory cytokines, which are signaling molecules that further amplify inflammation. This is similar to shouting for help in a crowded room—though the intention is to call for assistance, it may cause confusion and more chaos. As more cytokines are released, the inflammation intensifies, leading to even more oxidative damage and perpetuating a cycle that's hard to break.

Chronic oxidative stress not only wreaks havoc on the intestinal lining but can also disrupt the balance of microbial communities found in the gut, known as the gut microbiota (the collection of all the microorganisms that reside in the gut. A healthy gut microbiome (the entire habitat of these microorganisms, including the microorganisms, their collective genetic material, the surrounding environmental conditions, and the substances they produce) is essential for digestion, immune function, and even mental health. However, the increased oxidative stress can lead to a loss of beneficial bacteria and an overgrowth of harmful ones, creating a dysbiotic state. This imbalance can exacerbate inflammation and the symptoms typically associated with IBD, like diarrhea and abdominal pain. Think of the gut microbiota as a team of helpful workers; if some leave and others take their place that are unhelpful workers, the overall efficiency of the factory may suffer, leading to problems in production—in this case, proper digestion and nutrient absorption.

As these interconnected processes continue, they can result in serious complications such as ulceration of the intestinal lining and increased permeability of the gut. Ulcers can create painful sores and bleeding, while an increased permeability—leaky gut—allows even more harmful substances to enter the bloodstream. This condition creates a vicious cycle where the ongoing oxidative stress and inflammation not only exacerbate existing symptoms but can also lead to further damage to the intestines, making the management of IBD more challenging.

In summary, oxidative stress plays a critical role in the development and worsening of IBD. By damaging the intestinal lining, causing a misguided immune response, and altering the gut microbiota, oxidative stress fosters an environment where inflammation and injury can thrive. Ultimately, addressing oxidative stress could lead to better outcomes for those living with IBD, helping them reclaim their health and improve their quality of life.

Diabetes

Oxidative stress is a key player in the development and progression of diabetes, particularly type 2 diabetes. At the heart of this process are ROS that can damage various components of our cells. Normally, our body can neutralize these harmful compounds with antioxidants, but diabetes impairs this ability, leading to oxidative stress and a plethora of diabetes-related complications. In the context of diabetes, one of the critical ways that oxidative stress manifests is through the impairment of insulin signaling pathways.[52] When oxidative stress occurs, it signals that excessive damage is happening at the cellular level. One key group of proteins called serine/threonine kinases gets activated during this process. When these kinases are activated by oxidative stress, they can interfere with insulin signaling. Think of insulin as a key that helps open the doors to our cells so that sugar from the bloodstream can be used for energy. When oxidative stress is present, the actions of these kinases can cause confusion in the signaling process, making it harder for insulin to do its job. As muscle and fat tissues fail to respond effectively to insulin, excess sugar (glucose) remains in the blood, which can contribute to insulin resistance—a condition where the cells don't respond well to insulin, making it more difficult to manage blood sugar levels effectively.

The dysfunction caused by oxidative stress doesn't stop at insulin signaling. The pancreas, which produces insulin, contains beta cells that are particularly susceptible to oxidative damage.[53] When these cells are harmed, their ability to produce insulin diminishes, which only exacerbates the problem of high blood sugar. As blood glucose levels rise, this triggers further ROS production, creating a vicious cycle that makes managing diabetes increasingly challenging. This cycle of oxidative stress and insulin resistance can lead to a series of complications associated with diabetes, such as neuropathy (nerve damage), retinopathy (damage to the eyes), and cardiovascular disease (atherosclerosis, cardiomyopathy), increasing the risk of heart issues and strokes.[54] These are serious conditions that can significantly impact quality of life and require careful management.

Diabetes can change how certain protective enzymes work, including GPX and GSH, which are part of the body's main antioxidant defense system. Enzymes are proteins that speed up chemical reactions in the body, and when these antioxidant enzymes don't function properly, cells become less able to neutralize free radicals. As a result, cells are more vulnerable to oxidative stress and injury.

It also causes protein glycation—what happens when sugar attaches to proteins like a sticky residue, interfering with their normal function and accelerating damage over time. Oxidative stress in diabetes increases lipid peroxidation—a process where fats in the body are damaged by free radicals, further damaging enzymes, cellular machinery, and increasing insulin resistance. High blood sugar levels generate high levels of oxidative stress, which can react with certain amino acids (the building blocks of protein), producing modified, denatured (unfolded, misfolded, and damaged), or nonfunctional (unable to perform their responsibilities) proteins. This is a problem because proteins are essential molecules in the body that perform a wide variety of functions. They're involved in everything from building tissues and transporting nutrients to fighting infection and facilitating chemical reactions. When proteins are damaged or don't work correctly due to oxidative stress, it can disrupt these critical processes and contribute to the complications of diabetes. Another deleterious result of oxidative stress during diabetes is decreased levels of vitamin C in the body. This reduces the body's ability to combat oxidative stress, letting it run rampant in the body, damaging multiple cells and tissues.

Metabolic Syndrome

Metabolic syndrome is a cluster of related conditions that occur together, including abdominal obesity (increased waist circumference), high blood pressure (hypertension), unfavorable lipid profiles (high triglycerides, low HDL cholesterol), and insulin resistance (when the body's cells don't respond well to insulin). These conditions are all closely tied to oxidative stress. When the levels of ROS rise too high, they can disrupt normal metabolic processes, or the body's usual way of turning food into energy. This disruption contributes to insulin resistance, preventing glucose (sugar) from entering cells efficiently and leading to elevated blood sugar levels, further worsening metabolic syndrome.

Moreover, oxidative stress doesn't just impact insulin; it also triggers chronic inflammation in the body. When oxidative stress levels are elevated, it can lead to a state of ongoing inflammation, which adds to the problems associated with obesity. In simple terms, think of inflammation as the body's way of signaling that something is wrong. Chronic inflammation, especially with excess fat tissue, can worsen metabolic health. Excess fat cells release inflammatory substances that create a self-perpetuating cycle in which inflammation and oxidative stress reinforce one another. This further complicates metabolic syndrome, making it harder for the body to function effectively and increasing the risk of serious health issues over time.

Additionally, oxidative stress can interfere with lipid metabolism, which is the way the body processes fats. When this happens, it can lead to abnormal cholesterol levels, meaning that you might have high LDL cholesterol and triglycerides and low HDL cholesterol.[55] This is particularly concerning because the combination of high triglycerides and LDL cholesterol with low HDL is highly predictive of developing heart disease and other cardiovascular problems. It's like having a traffic jam in your arteries, which can lead to more severe issues down the road.

Bioactive signaling proteins (hormones) secreted by fat (adipose) tissue, called adipokines, serve as messengers that travel throughout the bloodstream, influencing a wide range of physiological processes. Your fat tissue isn't just storage; it's a busy factory producing hormones called adipokines that affect your whole body. Think of these adipokines as messengers. Some, like leptin, tell your brain you're hungry and can actually worsen oxidative stress by ramping up harmful ROS production and weakening your body's antioxidant defenses. Too much leptin can also stir up inflammation. On the flip side, adiponectin is a helpful messenger that improves sugar and fat handling, calms inflammation, and fights oxidative stress. It's like a peacekeeper, and having more of it is good. Then there's visfatin (also known as pre-B cell colony-enhancing factor or nicotinamide phosphoribosyltransferase), which acts more like a troublemaker, increasing inflammation and oxidative stress, and its levels rise with increased body fat. So, essentially, these fat-produced messengers play a major role in balancing, or unbalancing, the oxidative stress levels within your body.

Essentially, an overproduction of ROS, triggered by conditions like obesity, insulin resistance, and chronic inflammation, creates a state of oxidative stress that inflicts significant cellular damage.[56] This stress particularly targets lipids within cell membranes, leading to the formation of toxic lipid peroxides which overwhelm the body's natural detoxification systems. Similarly, proteins and nucleic acids are also vulnerable to ROS damage, resulting in dysfunctional proteins that accumulate and disrupt cellular processes, potentially triggering cell death. Furthermore, the disruption of redox signaling by excess ROS promotes pro-inflammatory and pro-fibrotic pathways, impairing insulin metabolic signaling and endothelial function, ultimately contributing to cardiovascular and renal complications. In short, the unchecked generation of ROS cascades into a series of detrimental effects, compromising cellular integrity and systemic health.

Cardiovascular Disease

Oxidative stress plays a significant role in cardiovascular disease, and its effects can create a series of harmful changes in the body that ultimately lead to serious health problems like heart attacks and strokes. Central to this issue is the damage

ROS causes to the cells lining our blood vessels, called endothelial cells. These cells are vital because they form a barrier between the bloodstream and the surrounding tissues, and they help regulate blood flow and pressure.

When oxidative stress is present, it damages endothelial cells (cells that line the interior surface of blood vessels, lymphatic vessels, and the heart), disrupting their normal function.[57] Oxidative stress activates endothelial cells, switching them from being smooth and helpful to sticky (prothrombotic) and inflammatory, which is the first step towards blood vessel damage and blood clots. One crucial job of healthy endothelial cells is to produce nitric oxide, a powerful molecule that helps blood vessels relax and widen, promoting healthy blood flow. Think of nitric oxide as a kind of lubricant that keeps blood flowing smoothly. When oxidative stress damages the endothelial cells and reduces nitric oxide production, it leads to something called endothelial dysfunction. This means that the blood vessels can't respond properly to signals that tell them to widen, resulting in increased arterial stiffness. This stiffness can contribute to elevated blood pressure, which is a significant risk factor for cardiovascular diseases.

Furthermore, oxidative stress also plays a role in how our body handles cholesterol, particularly low-density lipoprotein (LDL) cholesterol. Under normal conditions, LDL cholesterol travels through the bloodstream and delivers fats to cells where they are needed. However, when oxidative stress is high, ROS can oxidize LDL cholesterol, making it more likely to stick to the walls of arteries. This process leads to the formation of plaque, which is a buildup of cholesterol, fats, and other substances. Over time, these plaques narrow the arteries and reduce blood flow, like a clogged pipe that restricts water flow in a plumbing system.

This plaque formation can also trigger chronic inflammation in the body.[58] The immune system detects the oxidative damage and the presence of plaques, leading to an inflammatory response designed to heal the area. While inflammation is part of the body's protective mechanism, being in a state of chronic inflammation can worsen the problem. It can attract more immune cells to the area, continuing to promote plaque growth and further contributing to the narrowing of the arteries. This dangerous combination of endothelial dysfunction, increased blood pressure, and chronic inflammation creates a perfect storm for cardiovascular diseases. Ultimately, these interconnected processes significantly raise the risk of severe cardiovascular events, such as heart disease, heart attacks, and strokes.

Neurodegenerative Conditions (Alzheimer's and Parkinson's Disease)

Oxidative stress profoundly impacts the brain and is a major player in the development of neurodegenerative diseases like Alzheimer's and Parkinson's disease. In the context of neurodegenerative diseases, oxidative stress leads to damage in neurons, the specialized cells responsible for transmitting information throughout the brain and nervous system. When neurons suffer from oxidative damage, they can lose their ability to function properly, leading to serious cognitive and motor issues.

The brain consumes a large amount of oxygen and is rich in lipids, making it extremely susceptible to oxidative stress. Given the high metabolic activity of the brain and its substantial consumption of oxygen, significant ROS is produced as a byproduct, which must be neutralized by a highly efficient and active antioxidant system in the brain. The brain has several antioxidant systems that work to protect it against this ROS increase, including enzymatic antioxidants (SOD, CAT, GPX), non-enzymatic antioxidants (GSH, vitamin E, vitamin C), metal-binding proteins (ferritin, transferrin), the Nrf2 pathway, and glial cells (astrocytes, microglia). These antioxidant systems work synergistically to protect the brain from the damaging effects of ROS generated during oxygen consumption and other metabolic processes. Dysregulation or depletion of these antioxidant defenses can lead to oxidative stress, cell damage, and contribute to neurodegenerative diseases and other neurological disorders.

In Alzheimer's disease, the accumulation of ROS plays a significant role in the formation of amyloid-beta plaques and tau tangles.[59] These plaques and tangles are considered hallmarks of the disease and are associated with neuronal damage and death. The process begins when oxidative stress causes the misfolding and aggregation of proteins, including amyloid-beta. As these proteins accumulate outside neurons, they form plaques that disrupt communication between nerve cells. In addition, oxidative stress can cause tau proteins—critical for maintaining structure within neurons—to become hyperphosphorylated, meaning they acquire too many chemical phosphate tags that distort their shape and cause them to twist and clump together, leading to the formation of neurofibrillary tangles inside the cells. The presence of both plaques and tangles can severely impair neuronal function, resulting in the cognitive decline and memory loss characteristic of Alzheimer's disease.

The relationship between tau tangles, beta-amyloid plaques, and neurodegenerative diseases like Alzheimer's disease is complex and remains an area of active research. For many years, scientists believed that the buildup of amyloid-beta was the main trigger that damaged tau proteins and killed brain cells, but newer research shows the disease likely develops through multiple interacting processes rather than a single cause.

The traditional view posits that beta-amyloid plaques initiate a cascade of events leading to neurodegeneration. However, recent studies have suggested that the relationship may not be as straightforward as originally thought. Some research indicates that soluble forms of amyloid-beta, rather than the insoluble plaques, might be more closely associated with synaptic dysfunction and cognitive deficits. This suggests that merely targeting plaque removal may not be sufficient for treating or preventing Alzheimer's disease.

Tau tangles appear to correlate more strongly with neurodegeneration and cognitive decline than amyloid plaques in some studies. The hyperphosphorylated tau protein may disrupt cellular processes essential for neuron survival and function, which raises questions about the initial primacy of amyloid-beta in the disease's progression.

The role of neuroinflammation in Alzheimer's disease has gained attention in recent years. Chronic inflammation in the brain may exacerbate the neurodegenerative process and could be influenced by both amyloid-beta and tau pathology. This suggests that the immune response, rather than the presence of plaques and tangles alone, could be pivotal in the progression of the disease.

The notion that neurodegeneration, particularly Alzheimer's disease, may be related to a form of diabetes—referred to as "type 3 diabetes"—has garnered attention in recent years. This idea stems from observations that insulin resistance and glucose metabolism abnormalities in the brain could contribute to neurodegenerative processes. Indeed, the relationship between amyloid-beta/tau pathology and metabolic dysfunction is increasingly recognized as bidirectional and complex. For instance, amyloid-beta and tau accumulation can impair glucose metabolism in brain cells like astrocytes, affecting their ability to support neurons. Conversely, impaired insulin signaling and reduced glucose uptake can potentially exacerbate amyloid-beta and tau pathology by affecting clearance mechanisms and increasing oxidative stress. These newer findings don't necessarily negate the importance of amyloid-beta and tau but rather highlight the intricate network of pathological processes involved. It's likely that a multifaceted approach, targeting both the misfolded proteins and the metabolic dysfunctions, will be necessary for effective future treatments. Think of amyloid and tau as key players in a complex orchestra, and metabolic dysfunction as a condition that throws the entire orchestra out of tune. Understanding all the instruments and their interactions is crucial.

Insulin is well known for its role in regulating glucose metabolism in the body, but it is also crucial for brain function. It plays a role in synaptic plasticity (the ability of synapses to strengthen or weaken over time, which is essential for learning and memory) and helps regulate neuronal growth and survival. In individuals with insulin resistance or impaired insulin signaling, these processes can be disrupted, potentially accelerating neurodegeneration. The brain relies heavily on glucose as an energy source, and impaired glucose metabolism may lead to neuronal dysfunction and death. Early observations in people with Alzheimer's indicated reduced glucose metabolism in certain brain regions, particularly those associated with memory and cognitive functions. This has been interpreted as a possible sign of underlying neurodegeneration linked to metabolic issues akin to diabetes.

Oxidative stress can play a significant role in dysregulating glucose metabolism in the brain and may initiate an inflammatory cascade. The relationship between oxidative stress, glucose metabolism, and inflammation is complex and interconnected.[60] As discussed earlier, oxidative stress can disrupt insulin signaling, which could reduce glucose uptake and metabolism in neurons. High levels of oxidative stress can also damage mitochondria, which are crucial for energy production and glucose oxidation. Mitochondrial dysfunction can lead to inadequate energy supply for neurons, which may further exacerbate metabolic dysregulation.[61] The brain relies heavily on glucose as an energy source, so any disruption in glucose metabolism can threaten neuronal health and longevity.

Moreover, oxidative stress is strongly associated with inflammatory processes that may lead to an inflammatory brain environment. ROS can activate signaling pathways such as nuclear factor kappa B (NF-κB), damaging neurons. The damage caused by oxidative stress can provoke an immune response. Damaged cells and tissues release signals that attract immune cells, leading to chronic inflammation. Additionally, oxidative stress can activate glial cells (astrocytes and microglia), which are involved in the brain's immune response. When activated, these cells can release additional inflammatory mediators, perpetuating the inflammatory cycle. Chronic inflammation in the brain can further exacerbate oxidative stress and metabolic dysregulation. The bidirectional relationship means that inflammation can contribute to further neuronal damage and loss of metabolic control, potentially leading to neurodegenerative diseases, such as Alzheimer's disease. Ultimately, oxidative stress can act as a critical trigger that disrupts glucose metabolism in the brain, leading to further inflammatory responses.

Neurodegeneration may follow a cascading event that begins with insulin resistance, which impairs brain glucose metabolism and mitochondrial function, promotes the formation of tau proteins and plaques, and ultimately increases ROS production. There is growing evidence that insulin resistance in peripheral tissues can influence brain function, leading to a range of neurological issues, including cognitive decline and neurodegenerative diseases.[62] When insulin resistance occurs, glucose metabolism in the brain can be impaired. Impaired glucose metabolism in the brain can disrupt neuronal function and increase vulnerability to damage, which is linked to cognitive decline and neurodegeneration. Reduced glucose metabolism can further compromise mitochondrial function, which is a well-recognized feature of neurodegenerative diseases, including Alzheimer's and Parkinson's diseases. There is evidence suggesting that metabolic dysregulation, including that caused by insulin resistance and impaired glucose metabolism, can exacerbate the misfolding and aggregation of tau proteins and promote amyloid plaque formation.[63] The interplay of mitochondrial dysfunction, impaired glucose metabolism, and pro-inflammatory states can elevate ROS levels. Excess ROS can damage cellular components, promote inflammation, and further contribute to neurodegeneration.

Similarly, in Parkinson's disease, oxidative stress selectively damages dopamine-producing neurons, which are crucial for coordinating smooth and controlled movements.[64] These neurons reside in a part of the brain called the substantia nigra—a crucial brain region that primarily functions in the regulation of movement and coordination by producing the neurotransmitter dopamine. When ROS levels rise, it can lead to lipid peroxidation—the process by which free radicals attack fats in the neuronal membranes—resulting in cellular dysfunction and death. This is particularly concerning because dopamine is essential for regulating movement, mood, and cognition. As these neurons die, individuals may experience tremors, stiffness, and difficulty with balance and coordination, hallmark symptoms of Parkinson's disease. The combination of neuronal loss and the production of inflammatory substances further exacerbates the neurodegenerative process, creating a vicious cycle that accelerates disease progression.

Like Alzheimer's disease, recent research suggests that Parkinson's disease is correlated with metabolic dysfunction, especially impaired insulin sensitivity. Research indicates that the buildup of abnormal proteins called alpha-synuclein and the loss of nerve cells in the substantia nigra may be linked to problems with mitochondria caused by issues with fat and sugar processing, as well as insulin resistance.[65] This process might work by insulin receptor proteins connecting with other proteins that help keep nerve cells alive and are involved in the communication related to cell damage. In other words, disturbances in how the brain uses insulin might be involved in the death of dopamine-producing cells.

The interplay between oxidative stress, inflammation, and brain glucose metabolism is critical in unraveling the causes of both Alzheimer's and Parkinson's diseases. When neurons are damaged, they can release different signaling molecules that attract immune cells to the site of injury. While this immune response is intended to protect the brain, excessive activation can lead to neuroinflammation, which, in turn, can release more ROS, creating an even greater state of oxidative stress. This cycle not only perpetuates neuronal damage but also contributes to the progression of these diseases.

Liver and Kidney Disease

Oxidative stress plays a significant role in liver and kidney diseases by impairing organ function and contributing to tissue damage. In the liver, oxidative stress can cause serious problems. Due to its intense metabolic and detoxification activities, the liver inherently generates ROS as byproducts. Mitochondria within liver cells also contribute to ROS production during energy generation. Furthermore, immune responses within the liver involve ROS production. While some ROS is normal, excess production can lead to inflammation, which is the body's way of responding to injury or infection. Too much inflammation can be harmful and may pave the way for diseases like fatty liver disease, where fat builds up in liver cells, and cirrhosis, which is scarring of the liver tissue.[66] The oxidative damage to liver cells can disrupt their normal functions, which are essential for processing nutrients and filtering toxins from the blood. In extreme cases, this damage can trigger apoptosis, or programmed cell death, which further diminishes liver function and can lead to conditions like liver failure.

Similarly, in the kidneys, oxidative stress can be particularly damaging. The kidneys are responsible for filtering waste products from the blood and balancing fluids in the body. When oxidative stress comes into play, it can injure the tiny filtering units in the kidneys called glomeruli, as well as the kidney tubules that further process the filtrate.[67] This damage hampers the kidneys' ability to do their job, which can lead to the buildup of waste in the body and problems with blood pressure. Additionally, oxidative stress can trigger inflammatory responses in the kidneys, leading to fibrosis, which is the thickening and scarring of kidney tissue. Over time, this can exacerbate conditions like chronic kidney disease, making it harder for the kidneys to function properly. Several things can cause this, like high blood pressure, inflammation, and toxins that the kidneys normally filter out. When these things go wrong, the kidney cells produce too many ROS, which can lead to the kidneys not working as well as they should over time. The impact of oxidative stress on the liver and kidneys highlights how important it is to keep these organ systems in balance.

Fibromyalgia and Chronic Fatigue Syndrome

Fibromyalgia and chronic fatigue syndrome (CFS) are complex disorders that can significantly impact a person's quality of life. They often present with ongoing pain, extreme tiredness, and a variety of other symptoms, like sleep disturbances and mood issues. Recent studies suggest that oxidative stress plays a pivotal role in these conditions, and understanding this connection may support more targeted approaches to management.

In conditions like fibromyalgia and CFS, elevated ROS can lead to muscle injury and heightened sensitivity to pain. This can stem from a variety of factors, including certain molecules in the body that can lead to oxidative stress and inflammation.[68] One of these pro-oxidative factors is nitric oxide. While nitric oxide is a crucial molecule that helps regulate blood flow and can even have protective effects, when levels become too high, it can start having negative effects. In fibromyalgia, excess nitric oxide can contribute to pain by promoting inflammation. Think of it like a fire that's already burning; adding more fuel (in this case, nitric oxide) can cause the fire to grow, leading to more inflammation and a greater sensation of pain.

Another factor is lipid peroxidation.[69] This process involves the breakdown of fats in the body due to oxidative stress. When our cells experience oxidative damage, it can cause fats in the cell membranes to become unstable and break down. The breakdown products of these fats can signal pain and inflammation, intensifying feelings of discomfort. Essentially, when cell membranes are compromised due to lipid peroxidation, it can make them more susceptible to pain and inflammation.

Mitophagy is another term that may sound complex, but it's a natural process your body uses to clean up damaged mitochondria, the energy producers in our cells. In fibromyalgia, however, there can be either too much or too little mitophagy.[70] In some cases, if the body is not effectively clearing out these damaged mitochondria, it can lead to a buildup of toxins that cause further stress and inflammation. If this happens, it's like having an overflowing trash can in your house: the mess starts to build up and can create an uncomfortable environment. This inability to manage damaged mitochondria contributes to feelings of fatigue and pain.

These factors—increased levels of nitric oxide, lipid peroxidation, and issues with mitophagy—lead to inflammation and oxidative stress, which heighten pain sensitivity. Essentially, oxidative damage affects how nerve signals are processed, making the brain perceive pain more intensely than it normally would. This altered nerve signaling can result in widespread pain, which is a hallmark symptom of both fibromyalgia and CFS.

Evaluation of serum biomarkers associated with oxidative stress has identified significantly elevated levels of ROS, , malondialdehyde (MDA) and F2-isoprostane, and tumor necrosis factor-alpha (TNF-α) in people with CFS.[71] The researchers also noted that their antioxidant capacity was decreased, including total antioxidant activity, SOD, and GSH when compared to healthy subjects. These findings confirm that oxidative stress is a key contributor in the pathophysiology of CFS.

Another study looked at isoprostanes in people with CFS as a major biomarker of oxidative stress and clinical symptoms.[72] Isoprostanes are like damage markers that show up when your body's cells are under oxidative stress. As free radicals damage fats in your cell membranes, isoprostanes are produced. By measuring the levels of isoprostanes in someone's blood or urine, scientists can get an idea of how much oxidative stress their body is experiencing. Higher levels of isoprostanes generally indicate higher levels of oxidative stress. Those with active CFS had significantly elevated isoprostane levels and higher levels of oxidized LDL cholesterol.

A third study looked at zinc levels in the blood of people with CFS compared to healthy people.[73] They found that people with CFS had significantly lower zinc levels. Also, the lower the zinc, the worse their CFS symptoms and the more they felt like they had an infection. Low zinc was also linked to signs of inflammation in their blood and problems with their immune system's T-cells, which are important for fighting off infections. Since zinc helps protect against cell damage caused by oxidative stress, these findings suggest that CFS might involve increased cell damage. The researchers concluded that some people with CFS might benefit from taking zinc and other antioxidants. While other factors, such as immune function and hormone regulation, also contribute to CFS, the amount of oxidative stress experienced may be a leading indicator of symptom severity.

The lasting effects of oxidative stress can also lead to broader feelings of malaise. Many individuals report feeling unwell overall, which can stem from the body's ongoing battle with high levels of ROS. This cycle of pain and fatigue can be incredibly frustrating for those affected, as attempts to rest or recover often feel fruitless.

Chronic Inflammatory Conditions

A diversity of chronic inflammatory conditions are associated with oxidative stress.[74] Chronic inflammatory conditions, such as rheumatoid arthritis and lupus, can be incredibly challenging to live with, causing ongoing pain and fatigue. One of the key players in making these conditions worse is oxidative stress. To put it simply, oxidative stress can set off a chain reaction that fuels inflammation, further contributing to pain and other symptoms.

When ROS levels are high, they can activate immune cells, which are meant to protect the body from harmful invaders like bacteria or viruses. However, in chronic inflammatory conditions, these immune cells can become overactive and start attacking healthy tissues instead of just focusing on threats.[75] Imagine a security guard who gets so carried away doing their job that they start to see everyone as a potential threat, even people who are just going about their daily lives. This overactivity leads to persistent inflammation, which can worsen symptoms and lead to further damage in the joints and other organs.

Furthermore, oxidative stress impacts not only the immune cells but also the signals in the body that help regulate inflammation. In a healthy situation, the body has checks and balances in place to manage inflammation, making sure it doesn't go too far. However, when oxidative stress disrupts these regulatory systems, it can result in an uncontrolled inflammatory response. This means that even after an initial injury or threat has been managed, the inflammation can continue, causing prolonged pain and damage. Indeed, disease activity can directly relate to total oxidative stress in chronic inflammatory conditions.[76] Picture a fire burning out of control despite the fire department's efforts to contain it—without proper regulation, the damage keeps escalating.

The ongoing oxidative damage in the body can lead to significant challenges beyond inflammation. As tissues get damaged over time, they can no longer function properly, increasing pain and reducing quality of life. For instance, damaged cartilage in joints can lead to stiffness and limited mobility, while systemic inflammation from conditions like lupus can affect various organs, further complicating health. As a result, individuals may experience exhaustion and a decreased ability to enjoy regular activities, leading to a decline in overall well-being.

The role of oxidative stress in chronic inflammatory conditions may involve disruptions to mitochondrial function and mesenchymal stem cells. These multipotent stomal cells can become bone cells (osteoblasts), cartilage cells (chondrocytes), fat cells (adipocytes), or muscle cells (myocytes) and also function as signaling cells that secrete bioactive factors important for tissue regeneration.[77] An oxidative environment causes mesenchymal stem cells (MSCs) to become senescent, meaning the cells stop dividing and age prematurely. This causes the MSCs to decrease their mitochondrial membrane potential, reduce ATP production, and reduce their oxygen consumption rate. Stem cells are like the body's repair crew, helping to fix damaged tissues. Senescence is a process where cells age and stop working properly. So, senescent stem cells are like repair crew members who are too old to do their job; they can't fix the tissue damage. To make matters worse, these old, senescent stem cells don't just sit there doing nothing. They also cause trouble by releasing substances that trigger inflammation. Research indicates that oxidative stress interferes with the ability of MSCs to produce energy. This energy deficit leads to an increase in the production of ROS. High levels of ROS cause even more oxidative stress, which in turn causes the cells to become senescent. Senescent MSCs can no longer regulate immune responses and may further increase inflammation, contributing to the progression of chronic inflammatory conditions. Therefore, senescent stem cells contribute to tissue degeneration in two ways: they fail to repair the tissue, and they also trigger harmful inflammation.[78]

Hormone Imbalance

Oxidative stress not only affects our immune system and leads to chronic inflammation but can also play a significant role in disrupting hormone balance in both men and women. Hormones are the chemical messengers in our body that regulate everything from mood and metabolism to reproductive functions. Certain hormones, such as melatonin, insulin, estrogen, and progesterone, display antioxidant features, while others like thyroid hormones, corticosteroids, and catecholamines elicit the generation of free radicals.[79] When oxidative stress increases the levels of ROS, it can interfere with the normal processes that maintain hormonal balance, leading to a variety of health issues.

Interestingly, men tend to have a lower antioxidant capacity than women and are therefore more susceptible to oxidative stress.[80,81] In men, oxidative stress can particularly influence the relationship between testosterone and estrogen. Testosterone is a key hormone that supports many essential functions, including muscle mass, energy levels, sexual function, and mood stability. Under normal circumstances, some testosterone is converted into estrogen through a process called aromatization. However, when there are high levels of ROS due to oxidative stress, this conversion process can become dysregulated,[82] and higher oxidative stress levels are linked to lower testosterone levels.[83] Instead of maintaining healthy testosterone levels, the body may end up producing more estrogen than usual, leading to a state known as estrogen dominance. Oxidative stress is known to damage sperm and impair male reproductive function, and disruptions in the testosterone-estrogen balance due to altered aromatase activity could be a contributing factor. This imbalance can result in a series of symptoms, including increased body fat, especially around the abdomen, fatigue, mood swings, trouble concentrating, and decreased libido. Essentially, the stress on the body from oxidative damage can lead to hormonal turmoil that affects a man's overall health and quality of life.

Furthermore, inflammation, specifically proinflammatory cytokines, alter hormone balance. A plethora of research shows that aromatase activity and gene transcription are highly susceptible to proinflammatory cytokines.[84] These cytokines, such as TNF-α and IL-6, can directly upregulate aromatase expression in various tissues, including adipose tissue. Upregulation contributes to an increased conversion of testosterone to estradiol, potentially exacerbating conditions linked to estrogen dominance. Belly fat and low testosterone in men often go hand-in-hand. This is because abdominal fat, also known as visceral fat, is more than just stored energy; it's an active tissue that produces various hormones and inflammatory substances. One of the key enzymes found in this fat is aromatase, which converts testosterone into estrogen. So, when a man has excess abdominal fat, there's more aromatase present, leading to a greater conversion of testosterone to estrogen.[85,86] This results in lower testosterone levels and higher estrogen levels. Furthermore, the inflammatory environment created by abdominal fat can further stimulate aromatase activity. This hormonal imbalance can contribute to various health issues, including reduced muscle mass, decreased libido, and increased risk of metabolic diseases.

Despite a greater antioxidant capacity and the fact that estrogen regulates the expression of GSH and GPX in various tissues,[87,88,89] women are not immune to the effects of oxidative stress on hormone balance either. Just like in men, persistent oxidative stress can also exacerbate hormonal issues in women, particularly those related to estrogen. ROS can damage cellular components involved in hormone manufacturing and signaling.[90] Moreover, oxidative stress can affect the expression of genes involved in hormone production and metabolism.[91] In essence, oxidative stress can create a cascade of effects that disrupt the intricate hormonal balance in women, impacting reproductive health, sexual function, and overall well-being.

High levels of ROS can contribute to conditions like premenstrual syndrome (PMS), endometriosis, or polycystic ovary syndrome (PCOS), all of which are intrinsically linked to hormonal imbalances involving estrogen.[92] For instance, women with PCOS often experience higher levels of androgen hormones (which are male hormones that women also produce in smaller amounts), and oxidative stress may play a role in this imbalance. Elevated ROS can influence the ovaries' function, widening the gap between hormone production and the body's needs, leading to symptoms such as irregular menstrual cycles, weight gain, acne, and excess hair growth. Not surprisingly, oxidative stress can influence fertility.

Studies indicate that oxidative stress can interfere with follicular growth, oocyte maturation, and the development of the endometrium, all of which are crucial for successful reproduction.[93] Oxidative stress can even increase the risk of pregnancy complication, such as preeclampsia and miscarriage.[94] This interplay between oxidative stress and hormone levels contributes significantly to the discomfort and health challenges women may face.

By addressing oxidative stress, individuals can potentially improve their hormone health, leading to a reduction in symptoms and a better quality of life overall. This approach highlights the importance of viewing the body as an interconnected system—taking care of one aspect, like oxidative health, can lead to improvements in another, such as hormonal balance. Ultimately, acknowledging the role of oxidative stress opens new avenues for managing hormone-related conditions, allowing both men and women to reclaim their health and vitality.

Skin Aging

Oxidative stress is one of the major culprits behind skin aging, affecting not only how our skin looks but also how it functions. The skin is the body's largest organ and first line of defense against various environmental stressors, such as ultraviolet (UV) radiation from the sun, pollution from vehicle emissions, and even toxins in our daily surroundings. When the skin is exposed to these agents, it produces ROS that can cause significant damage to skin cells.[95] This process is similar to how rust can form on metal when it's exposed to moisture and air over time—it's a slow, but cumulative effect that leaves the skin looking older and less vibrant.

One of the critical components affected by oxidative stress is collagen, a protein that provides structure and firmness to the skin, and elastin, which gives the skin its ability to stretch and bounce back. With ongoing exposure to oxidative stress, these proteins can become damaged and degrade, leading to a loss of skin's elasticity and firmness. As a result, you may start to notice wrinkles, fine lines, and sagging skin. Visualize a rubber band that you stretch repeatedly—it eventually loses its shape and no longer holds its tension. Similarly, the skin loses its youthful appearance as these proteins break down due to oxidative damage, making it appear less plump and more aged.

In addition to affecting the physical structure of the skin, oxidative stress can also disrupt its function by impairing the skin's barrier. The skin barrier serves as a protective shield, preventing moisture loss and keeping harmful bacteria and pollutants at bay. When oxidative stress damages the skin's barrier, it becomes less effective at retaining moisture, leading to dryness and an increased risk of irritation and sensitivity. This may also result in conditions like eczema or dermatitis, where the skin reacts negatively to environmental factors. Picture a house with broken windows; the elements can easily invade, making it harder to keep the inside of the house warm and comfortable. Similarly, a compromised skin barrier means vulnerability to external influences.

To counteract skin aging caused by oxidative stress, incorporating antioxidants into your skincare routine can be highly beneficial. Antioxidants are compounds that neutralize ROS, helping to minimize the damage they cause. Ingredients such as vitamin C, vitamin E, and green tea extract are examples of powerful antioxidants that can be found in serums and moisturizers. Using products rich in these nutrients can help protect the skin from further oxidative damage, promoting a healthier, more youthful complexion.

The relationship between oxidative stress and health conditions is a complex interplay, where cause and effect often blur together. Rather than viewing oxidative stress as merely a precursor to health issues, it's essential to recognize that it can be both a contributory factor and a consequence of various conditions, creating a vicious cycle within the body. For instance, prolonged oxidative stress can lead to chronic inflammation, which in turn may exacerbate conditions such as cardiovascular disease, diabetes, or neurodegenerative disorders. Conversely, these health issues can also generate further oxidative stress, perpetuating a cycle that can be difficult to escape.

Given the significant role that oxidative stress plays in a variety of ailments, it becomes vital to focus on strategies to maintain a balanced oxidative state. This involves managing free radicals through adequate intake of antioxidants, reducing exposure to oxidative stress-causing agents, managing stress, and taking targeted supplements. Antioxidants work by neutralizing free radicals, effectively breaking the cycle of damage that can lead to serious health conditions. When the body's natural antioxidant defense systems become overwhelmed—due to factors such as poor diet, pollution, smoking, or chronic stress—it can result in tissue damage that not only accelerates aging but also triggers the onset of various diseases.

Moreover, understanding the importance of antioxidants goes beyond mere dietary intake; it's also about adopting a holistic approach to health. Lifestyle choices such as regular exercise, good sleep hygiene, and stress management play a crucial role in enhancing the body's capacity to combat oxidative stress. By cultivating a lifestyle that prioritizes antioxidant-rich foods and healthy habits, individuals can improve their overall resilience against oxidative damage. In doing so, they not only mitigate the immediate impacts of oxidative stress but also promote long-term health and well-being. Ultimately, fostering a proactive approach to managing oxidative stress can empower us to take control of our health, breaking the cycle of damage and enabling a more vibrant and balanced life.

CHAPTER FOUR

Key Lifestyle Factors Influencing Oxidative Stress

The truth is, the same lifestyle behaviors that support general health can also help manage oxidative stress. You need to provide your body daily nutrients, engage in appropriate physical activity, get adequate sleep, hydrate, minimize exposure to toxins, manage stress and negative emotions, and limit or avoid the use of harmful substances.

Nutrition

A balanced diet that includes both plant-based and animal products is ideal for managing oxidative stress effectively. Nutrition plays a pivotal role in combating oxidative stress, as foods rich in antioxidants can help neutralize free radicals and mitigate oxidative damage in our bodies. Incorporating a diverse array of antioxidant-rich foods—such as colorful fruits and vegetables high in vitamins C and E, beta-carotene, and polyphenols—as well as nutrient-rich animal products—fatty fish, eggs, grass-fed and finished pasture-raised beef, and organ meats—can provide the necessary support for maintaining an optimal oxidative balance. Regular consumption of a variety of healthy foods, including berries, grass-fed beef, fatty fish, and nuts, contributes to a comprehensive approach to reducing oxidative stress and promoting overall health.

Nutrition is a fundamental element in combatting oxidative stress, and a diverse diet that includes both plant-based and animal products offers a wide range of antioxidants and nutrients. Berries provide antioxidants like anthocyanins, vitamin C, and flavonoids. Regular consumption of berries has been linked to improved cognitive function and reduced risk of chronic diseases.[96,97] Apples are a good source of quercetin and catechin, while grapes provide resveratrol, flavonoids, and vitamin C. A rich source of vitamin C, oranges also provide carotenoids and flavonoids. Broccoli, Brussels sprouts, and cauliflower contain glucosinolates, which can help the body detoxify and protect against oxidative stress. Nuts are not only a source of healthy fats but also provide vitamin E, which helps stabilize cell membranes and neutralize free radicals. Additionally, they contain compounds that support heart health and reduce inflammation. Chia and flaxseeds are high in omega-3 fatty acids and fiber, which can improve gut health and contribute to a balanced antioxidant intake. Eggs, particularly the yolks are an excellent source of the antioxidants lutein and zeaxanthin. Lean meats like grassfed beef and free-range organic chicken provide CoQ10 and other bioactive compounds that contribute to antioxidant status. Fatty fish such as salmon, mackerel, and sardines are high in omega-3 fatty acids, which have anti-inflammatory properties. Additionally, these fish are rich in carotenoids and astaxanthin, potent antioxidants that can help protect against oxidative stress. Nutrient-dense organ meats supply vitamins A, D, E, and essential minerals like selenium. These dietary components work synergistically to support the body's natural defenses against oxidative damage.

When comparing animal products like eggs and grass-fed beef to fruits and vegetables, it's essential to consider the different types of nutrients they provide and their bioavailability. Both food categories play crucial roles in a balanced diet, but they offer distinct nutritional profiles.

Nutrients Found in Animal Products

Found predominantly in animal products, vitamin B12 is essential for red blood cell formation, nerve function, and DNA synthesis. Plant-based products do not provide B12, making it critical for those following vegetarian or vegan diets to seek fortified foods or supplements.

Animal products, particularly fatty fish and egg yolks are rich in vitamin D3 (cholecalciferol), which is vital for bone health and calcium regulation. It also supports the immune system, keeping it strong to fight off infections. Furthermore, vitamin D contributes to healthy muscle function and may play a role in brain health. In fact, vitamin D is involved in the regulation of thousands of genes, potentially influencing—directly or indirectly—up to 5% of the human genome.[98,99,100]

Found in meat and fish, heme iron is absorbed more efficiently by the body compared to non-heme iron from plant sources. This is particularly important in preventing iron deficiency anemia. Iron from heme sources supports numerous bodily functions, including energy production, DNA synthesis, immune function, brain development and growth in children, and the production of certain hormones and cells.

Primarily found in meat, seafood, and dairy products, zinc is critical for immune function, wound healing, and DNA synthesis. While zinc is present in some plant foods, it is often less bioavailable due to phytates that inhibit absorption.

Grass-fed beef and certain animal products provide omega-3 fatty acids, particularly when the animals are pasture-raised. These fatty acids are crucial for heart and brain health. Grass-fed beef is typically 20%–40% saturated fat, around 50% unsaturated fat, and 5%–10% polyunsaturated fat.

Animal products contain all essential amino acids in sufficient quantities, making them complete, high-quality protein sources. Eggs, in particular, are often referred to as the gold standard for protein quality due to their balanced amino acid profile.

Nutrients Found in Fruits and Vegetables

Fruits and vegetables, especially citrus fruits, bell peppers, and leafy greens, are excellent sources of vitamin C, which supports immune function and acts as a powerful antioxidant.

Found abundantly in dark leafy greens and certain fruits (like oranges and bananas), folate is crucial for DNA synthesis and cellular division, and is particularly important during pregnancy to prevent neural tube defects.

Fruits and vegetables are rich in carotenoids (e.g., beta-carotene), which convert to vitamin A in the body and are important for vision, immune function, and skin health. However, the efficiency of this conversion can vary significantly among individuals—from 7.8% to 75%—based on several factors, including dietary factors, genetic factors, and physiological conditions.[101,102,103] It's estimated that for every 12 to 24 mcg of beta-carotene a person consumes, 1 mcg of retinol (active vitamin A) is produced. People with hypothyroidism, liver disease, malabsorption syndromes (e.g., celiac diseases, Crohn's disease), and genetic polymorphisms (BCMO1) can experience impaired conversion of beta-carotene to retinol.[104,105,106,107]

Fruits and vegetables are high in dietary fiber, which aids digestion, helps regulate blood sugar levels, and contributes to satiety, making them beneficial for weight management and gut health.

Fruits and vegetables contain various phytochemicals, such as flavonoids and polyphenols, which have been linked to reduced inflammation and lower risks of chronic diseases, such as heart disease and cancer.

Bioavailability Considerations

Bioavailability has been references a few times; this term refers to the degree and rate at which a nutrient or bioactive compound is absorbed and utilized by the body. Notable nutritional differences between animal and plant foods exist due to their inherent properties.

Nutrients in animal products like vitamin B12 and heme iron are absorbed more efficiently than their plant-based counterparts. For example, heme iron from animal sources can be absorbed at rates of 15-35%, while non-heme iron from plants is generally absorbed at rates of 2-20%.[108]

Fruits and vegetables often contain compounds that can enhance or inhibit the absorption of other nutrients. For example, vitamin C can enhance non-heme iron absorption, which can be beneficial for individuals relying on plant-based iron sources. On the other hand, plants are full of anti-nutrients. Anti-nutrients are natural compounds found in various plant-based foods that can interfere with the absorption of essential nutrients. Substances like phytates, oxalates, and lectins, while serving protective functions for the plants themselves, can bind to minerals like iron, zinc, and calcium, reducing their bioavailability in the human digestive system.[109] However, it's important to note that many plant foods containing anti-nutrients also offer a wealth of beneficial nutrients, so balance is key. Furthermore, techniques like soaking, sprouting, fermenting, and cooking can significantly reduce the anti-nutrient content, allowing for better nutrient absorption.

While it is clear that both plant-based and animal-based foods have their unique contributions to antioxidant intake, achieving a balance is crucial for overall health. Focusing solely on one food group may lead to deficiencies in essential nutrients that are vital for the body's ability to combat oxidative stress.

Vitamin B12 is a crucial nutrient for maintaining the body's antioxidant defenses and managing oxidative stress. Sufficient levels of vitamin B12 help maintain optimal levels of glutathione, enhancing the cell's ability to detoxify free radicals and reactive oxygen species (ROS). This action reduces oxidative damage to lipids, proteins, and DNA. By supporting the synthesis of glutathione, promoting healthy red blood cell formation, aiding in neurological health, and facilitating immune function, B12 contributes significantly to reducing oxidative damage. Adequate dietary intake of vitamin B12 is essential to ensure these protective functions are preserved, thereby helping the body effectively combat oxidative stress and its associated health issues.

One important consideration for those following a vegan diet is ensuring adequate intake of nutrients that may be limited without supplementation, such as B12, as deficiencies can affect brain health. A study led by researchers at the University of Oxford found that lower levels of vitamin B12 were associated with greater brain volume loss in older adults.[110] The research indicated that individuals with lower vitamin B12 levels experienced more significant brain shrinkage over a five-year period compared to those with higher levels. Obviously, this is less than ideal, as brain atrophy often leads to cognitive impairment.

A newborn's vitamin B12 storage exclusively comes from placental transfer during gestation and from animal foods after birth. Similarly, a case study reported cerebral atrophy in an infant exclusively breastfed by a vegetarian mother, highlighting the potential risk of vitamin B12 deficiency in such scenarios. The infant presented with neurological symptoms attributed to low vitamin B12 levels, emphasizing the importance of adequate B12 intake during pregnancy and breastfeeding.[111] Other research highlights failure to thrive, psychomotor regression—a loss of previously acquired motor skills, such as rolling, sitting, or crawling—anemia, and brain atrophy in infants breastfed by vegetarian mothers.[112,113,114] The collective evidence underscores the critical necessity of B12 supplementation for vegetarian mothers and their infants to avert severe and potentially irreversible developmental impairments.

In summary, animal products such as eggs and grass-fed beef provide essential nutrients that are often more bioavailable than those found in fruits and vegetables, including vitamins B12 and D3, heme iron, and high-quality protein. However, fruits and vegetables are vital for providing antioxidants, vitamins C and A, folate, and a range of phytochemicals that contribute to overall health. A balanced diet that incorporates both animal and plant foods can help ensure that individuals receive a comprehensive array of nutrients essential for good health. The synergistic effect of combining different antioxidant sources yields a more significant health benefit than relying on one type alone. The various phytochemicals and nutrients in a well-rounded diet enhance the body's resilience against oxidative damage.

Physical Activity

Engaging in physical activity is an integral part of a healthy lifestyle and can play a pivotal role in combating oxidative stress. Exercise, when performed in moderation and appropriate for individual health conditions and capabilities, enhances the body's antioxidant defense system by stimulating the production of endogenous antioxidant enzymes. These enzymes help neutralize free radicals, mitigating the effects of oxidative stress that can contribute to chronic health conditions and aging. However, it's essential to acknowledge that excessive exercise—especially without adequate recovery—can lead to an overproduction of free radicals and result in increased oxidative stress.

Moderate-intensity exercise has been shown to activate various cellular pathways that promote the synthesis and activity of antioxidants within the body.[115] This includes the key enzymes SOD, CAT, and GPX, which are integral to detoxifying ROS. Regularly engaging in physical activity can upregulate these antioxidant enzymes, fortifying the body's defenses against oxidative stress.

Research has demonstrated that individuals who engage in regular physical activity tend to have higher levels of antioxidants in their bloodstream compared to sedentary individuals.[116] This boost in antioxidant capacity can significantly lower the risk of oxidative stress-related diseases, including cardiovascular diseases, neurodegenerative disorders, and certain cancers.

Regular physical activity promotes overall metabolic health. It enhances insulin sensitivity, aids in weight management, and helps regulate blood sugar levels.[117,118] Improved metabolic health is closely linked to a reduced risk of oxidative stress. Insulin resistance and elevated blood sugar levels are associated with increased oxidative stress, while physical activity helps counteract these detrimental effects.

Additionally, exercise promotes healthy blood circulation, which is essential for transporting nutrients and oxygen to cells and facilitating the removal of metabolic waste products. Improved blood flow also enhances the delivery of antioxidants to tissues, further combating oxidative stress.

Physical activity improves cardiovascular function by strengthening the heart and improving the efficiency of the circulatory system. A healthy cardiovascular system is crucial for maintaining oxidative balance, as reduced blood flow and increased blood pressure are often associated with heightened oxidative stress.

Exercise has been linked to improvements in endothelial function—an essential component of vascular health.[119] A well-functioning endothelium helps maintain proper vascular tone and reduces the risk of atherosclerosis, also known as the hardening of arteries, which is further exacerbated by oxidative stress.

The benefits of physical activity extend beyond physical health; exercise has profound effects on mental health as well. Regular physical activity has been shown to enhance mood, reduce anxiety, and improve overall psychological well-being.[120] The release of endorphins during exercise promotes a sense of happiness and relaxation, which is crucial for maintaining a high quality of life, especially for older adults. Indeed, regular exercise is comparable or superior in efficacy to SSRI antidepressants for mild to moderate depression.[121]

Additionally, mental well-being is intricately linked to physical health. Chronic stress and poor mental health have been associated with increased oxidative stress, so maintaining a positive mental state can contribute to better overall oxidative balance.

While regular exercise has numerous benefits for combating oxidative stress, engaging in excessive exercise can have the opposite effect. Over-exercising can lead to an increase in oxidative stress, ultimately harming the body rather than helping it. Essentially, the intensity and duration of physical activity determines the levels of oxidative stress in the body.[122] This is particularly true for people with pre-existing conditions such as autoimmune disorders or thyroid dysfunction, since they likely already have impaired antioxidant defenses and the free radicals produced during prolonged exercise can contribute to and worsen inflammation, potentially triggering flares or exacerbating symptoms.

Intense and prolonged exercise can result in excessive free radical production. During vigorous exercise, the body experiences heightened metabolic activity and oxygen consumption, both of which are associated with increased ROS production. When the production of free radicals exceeds the body's ability to neutralize them with antioxidants, oxidative stress can ensue.

One of the crucial aspects of any exercise regimen is the recovery period. Inadequate recovery time can lead to cumulative fatigue and exacerbate oxidative stress. Muscles and tissues require time to repair and regenerate after intense workouts, and neglecting proper rest and recovery can impede this process. Without sufficient recovery, the body's antioxidant mechanisms may become overwhelmed, resulting in increased oxidative damage.

Over-exercising can also lead to hormonal disruptions. For example, excessive exercise can lower levels of important hormones like testosterone and negatively impact cortisol regulation.[123] Cortisol, a stress hormone, can increase oxidative stress when chronically elevated. Hormonal imbalances can further exacerbate fatigue, decreased recovery capacity, and increased susceptibility to oxidative damage.

It is essential to be aware of the signs and symptoms of over-exercising to prevent potential harm to your body. Here are some indicators that may suggest you are pushing your body beyond its limits:

- *Fatigue.* Persistent fatigue that does not improve with rest or sleep is a significant sign of overtraining. While some fatigue is normal after intense workouts, chronic exhaustion can indicate overstress on the body.

- *Increased Injuries.* Recurrent injuries, including strains, sprains, or chronic pain, may indicate that your body is not getting adequate rest and recovery time.

- *Mood Changes.* If you notice fluctuations in mood, increased irritability, anxiety, or depression that seem to coincide with your exercise routine, this could be a sign that your body is overwhelmed by stress.

- *Sleep Disturbances.* Difficulty sleeping or changes in sleep patterns can result from overtraining, as an overactive body and mind often struggle to relax.

- *Altered Appetite.* A sudden increase or decrease in appetite, or cravings for unhealthy foods, may indicate hormonal changes due to excessive exercise.

- *Weakened Immune Function.* If you find yourself getting sick more often than usual, it could be a sign that your body's immune defenses are compromised due to prolonged physical exertion without sufficient recovery.

To reap the numerous health benefits of exercise while minimizing the risks of oxidative stress, it's crucial to find the right balance. Pay attention to your body's signals. If you are feeling excessively fatigued or experiencing any of the symptoms of over-exercising, take a step back and allow yourself to rest. Schedule regular rest days into your exercise routine. Recovery is essential for muscle repair and maintaining optimal physical health. Incorporate a mix of low, moderate, and high-intensity workouts into your regimen to prevent burnout and allow your body to adapt and recover appropriately. Choose activities that you enjoy. Exercise should not feel like a chore; when you enjoy your workouts, you are more likely to stick with them. If you are unsure of how to design a balanced exercise program that aligns with your fitness level and health conditions, consider consulting a fitness professional or a healthcare provider.

Physical activity, when approached mindfully and tailored to individual needs, is a powerful ally in the fight against oxidative stress. The benefits of moderate exercise are extensive, from enhancing the body's antioxidant capacities to improving overall metabolic and cardiovascular health. However, it is crucial to recognize the potential pitfalls of over-exercising, which can lead to heightened oxidative stress and a cascade of health issues. Understanding the signs of over-exercising and finding a balance that works for your body is essential for optimizing health outcomes and maintaining a high quality of life. By cultivating a sustainable exercise regimen, you can enjoy the protective benefits of physical activity while safeguarding yourself against the potential downsides of excessive exertion.

Sleep

Sleep, often overlooked in conversations about health and wellness, plays a critical role in maintaining the body's ability to combat oxidative stress. During sleep, the body undergoes a range of restorative processes that facilitate not only physical recovery but also neurological maintenance and metabolic equilibrium. As oxidative stress is linked to various health issues, including chronic diseases and accelerated aging, understanding the interplay between sleep and oxidative stress is essential.

The body has a complex network of antioxidant defenses that work to mitigate oxidative damage. During sleep, certain endogenous antioxidants become particularly active, enhancing the body's defense mechanisms against oxidative stress.[124] When sleep is deprived, the body experiences an imbalance, leading to reduced levels of key antioxidants like GSH and CAT, particularly in the liver. This increases oxidative stress, which can damage cells and impair overall health. However, when sleep is restored, antioxidant levels recover, and the body's defense mechanisms are enhanced, helping to repair damage and restore balance. This underscores the importance of sleep in preventing oxidative damage and supporting long-term health.

In fact, research highlights the crucial role of sleep in protecting and repairing our cells, particularly in reducing oxidative stress and restoring mitochondria.[125] During wakefulness, our bodies generate energy at high rates, leading to the buildup of oxidative stress. If not addressed, this can damage our cells and accelerate aging. Sleep acts as a repair phase, allowing antioxidants to neutralize free radicals, restoring balance, and promoting the fusion and repair of mitochondria, the powerhouses of our cells. This process is essential for maintaining energy production, supporting brain function, and preventing long-term damage. The study suggests that adequate sleep is key to keeping our mitochondria healthy, which in turn supports overall well-being and longevity.

The primary antioxidant enzymes that become more active during sleep include:

- *SOD*. This enzyme converts superoxide, a harmful free radical, into hydrogen peroxide. Hydrogen peroxide is then further neutralized by other antioxidant systems. High levels of SOD during sleep can help reduce oxidative stress effectively.
- *CAT*. Catalase plays a vital role in breaking down hydrogen peroxide into water and oxygen, thereby protecting cells from oxidative damage. Sleep

enhances the activity of this enzyme, allowing the body to manage oxidative stress more effectively.

- *GPX*. Another vital antioxidant enzyme, glutathione peroxidase uses glutathione to reduce hydrogen peroxide and lipid peroxides. The synthesis of glutathione is boosted during sleep, making this enzyme particularly effective at neutralizing harmful oxidants when the body is at rest.

- *Endogenous Antioxidant Production*. Sleep facilitates the production of several key antioxidants, including glutathione, an important nonenzymatic antioxidant that is vital for detoxification. Glutathione levels naturally rise during sleep due to better availability of nutrients and lower metabolic demand during this restful state. This increase in glutathione levels helps in neutralizing free radicals generated through metabolic processes and environmental exposures.

The quality and duration of sleep are crucial for maximizing antioxidant activity. Poor sleep—whether due to insomnia, sleep apnea, or insufficient sleep—can lead to disruptions in the body's ability to regulate oxidative stress effectively. Research indicates that individuals who experience chronic sleep deprivation have elevated markers of oxidative stress and inflammation.[126]

High-quality sleep, characterized by uninterrupted cycles of rapid eye movement (REM) and non-REM sleep, is essential for an optimal antioxidant response. During REM sleep—the dreaming phase, your brain is very active. One important thing happening during this time is that your brain uses a special system called the glymphatic system. Think of this system like a cleaning crew that clears out the trash and waste products from your brain. When you're in REM sleep, the glymphatic system becomes more active, helping to remove waste substances that build up during the day. These waste products can cause damage to brain cells if they stay around, leading to oxidative stress, which can harm your brain over time. So, REM sleep helps keep your brain healthy by cleaning out these harmful waste products, allowing your brain to rest and recharge properly. Studies have shown that poor sleep quality can diminish the body's antioxidant capacity, leaving it more vulnerable to oxidative damage.[127,128]

The National Sleep Foundation recommends 7 to 9 hours of sleep per night for most adults. Getting less than this recommended amount can negatively impact the production and activity of antioxidants, increasing susceptibility to oxidative stress and its associated health risks.

Furthermore, sleep plays a pivotal role in mental health, which has a direct connection to oxidative stress. Chronic stress can lead to increased levels of cortisol, a hormone that can promote oxidative stress and inflammation. Sleep helps to regulate cortisol levels, and a well-rested body is better equipped to manage stress. By maintaining low stress levels through adequate sleep, individuals can effectively reduce the load of oxidative stress on their bodies.

To harness the benefits of sleep and combat oxidative stress, adopting healthy sleep habits is crucial. Here are some strategies for improving sleep quality:

- *Establishing a Consistent Sleep Schedule.* Going to bed and waking up at the same time every day helps regulate the body's internal clock, improving the quality of sleep.
- *Creating a Relaxing Sleep Environment.* A dark, cool, and quiet bedroom can enhance sleep quality. Consider using blackout curtains and relaxation techniques such as deep breathing or meditation before bedtime.
- *Limiting Screen Time.* The blue light emitted by screens can interfere with the production of melatonin, a hormone important for sleep. Limiting screen time in the evening—preferably one hour before bedtime—can promote better sleep quality.
- *Avoiding Stimulants.* Limiting the intake of caffeine and nicotine, particularly in the hours leading up to bedtime, can facilitate better sleep.
- *Physical Activity.* Regular physical activity has been shown to improve sleep quality. Engaging in moderate exercise can help you fall asleep faster and enjoy deeper sleep, although it's best to avoid vigorous activity too close to bedtime.

Sleep is not merely a period of rest but an active state that plays a vital role in maintaining the body's resilience against oxidative stress. By supporting the activity of key antioxidant enzymes and enhancing the production of crucial antioxidants such as glutathione, sleep serves as a powerful ally in the quest for health and well-being. Recognizing the importance of quality sleep and adopting healthy sleep habits is essential for anyone looking to combat oxidative stress and promote overall health. Investing in sleep is investing in long-term health, resilience, and a higher quality of life.

Hydration

Hydration is vital for maintaining optimal health and functioning, and its importance extends well beyond mere thirst quenching. Adequate hydration plays a crucial role in combating oxidative stress. Water facilitates numerous biological processes, including detoxification, nutrient transportation, and metabolic functions. Understanding the interconnection between hydration, oxidative stress, electrolytes, and pH (potential of hydrogen) can help maximize health outcomes and enhance cellular function.

Dehydration can exacerbate oxidative stress by impairing the body's ability to produce and utilize antioxidants effectively. Water is essential for numerous physiological functions, including:

- *Detoxification.* Proper hydration ensures that the kidneys can efficiently remove toxins and waste products from the body. When dehydrated, the kidneys conserve water, which can result in concentrated urine that may not effectively eliminate harmful substances, increasing oxidative stress.
- *Supporting Antioxidant Production.* Many antioxidants rely on water-soluble co-factors to function, including vitamins like vitamin C and minerals such as zinc and magnesium. Adequate hydration ensures that these nutrients are readily available to the cells where they are needed.
- *Maintaining Cellular Integrity.* Cells are composed largely of water (accounting for 70% or more of total cell mass), and hydration is crucial for maintaining cellular structure and function. When cells are dehydrated, they may lose their ability to produce energy effectively, leading to increased oxidative stress and diminished antioxidant defenses.
- *Regulating Temperature.* Water regulates body temperature through perspiration. Elevated temperatures can increase oxidative stress, so maintaining hydration helps ensure that body temperature remains within a healthy range. Specifically, elevated body temperature impairs mitochondrial function, leading to uncoupling of the electron transport chain and forming free radicals, increases the activity of enzymes that produce free radicals (NADPH oxidases and xanthine oxidase), reduces blood flow to visceral organs as blood flow is prioritized to the skin for cooling, and activates immune cells that in turn produce free radicals.

Electrolytes—such as sodium, potassium, calcium, and magnesium—are essential minerals that play a significant role in cellular function and overall hydration. They help regulate fluid balance, muscle contractions, and nerve signaling. Understanding the relationship between electrolytes and hydration is critical for combatting oxidative stress.

Electrolytes help maintain the osmotic balance within cells, allowing water to enter and remain within cells. Osmotic balance within cells refers to the regulation of water movement into and out of cells to maintain an appropriate concentration of dissolved substances (solutes) inside and outside the cell. This balance is crucial for proper cell function and overall hydration. Insufficient levels of electrolytes can lead to cellular dehydration, impairing the cells' ability to produce energy and respond to oxidative stress.

Electrolytes like potassium and sodium are crucial for proper nerve signaling. When dehydration occurs, electrolyte levels can become imbalanced, which may negatively impact muscle contraction and cognitive function, leading to further oxidative stress.

During exercise or physical activity, the body loses fluids and electrolytes through sweat. Replenishing these electrolytes is important for muscle recovery and performance, as well as reducing oxidative stress generated by rigorous physical activity.

The pH level of body fluids influences hydration and cellular function. The body relies on a tightly regulated pH for optimal biochemical processes, and deviations can disrupt cellular hydration and lead to oxidative stress.

The body works to maintain a slightly alkaline blood pH within a narrow range of 7.35 to 7.45 (average of 7.4). When the blood becomes too acidic (acidosis), it can lead to cellular dehydration, metabolic dysfunction, and increased ROS production. A proper pH balance helps facilitate efficient cellular metabolism, where antioxidants can effectively combat oxidative stress.

Indeed, drinking alkaline water to change your body's pH is largely pointless because of your body's built-in systems to maintain that super-tight pH balance in your blood. When you drink alkaline water, it first hits your stomach, which is highly acidic (pH 1.5-3.5) to digest food, and this acid quickly neutralizes any alkalinity from the water. Even if a tiny bit gets past, your lungs and kidneys constantly work to adjust your blood's pH, either by breathing out carbon dioxide (an acid) or by adjusting how much acid or base they release in your urine. These natural buffering systems are so powerful that they quickly bring your blood pH back to its normal, slightly alkaline range, regardless of what you drink.

The solubility of nutrients and minerals in the body is pH-dependent. Proper pH levels enhance the solubility of electrolytes, facilitating their absorption and transport into cells. This ensures that water, along with nutrients and electrolytes, can efficiently move across cell membranes, promoting cellular hydration and function.

Proper pH levels are essential for the function of many antioxidants. Glutathione, a crucial antioxidant, works inside your cells, not typically in the bloodstream. Each part of a cell, like the cytoplasm or mitochondria, has its own slightly different pH that's ideal for the enzymes and molecules working there. For example, glutathione, a key intracellular antioxidant, relies on appropriate

73

pH levels (range 4–8; ideal range of 6–7) for its optimum activity.[129,130] When pH levels are within the right range, antioxidants are more effective at neutralizing reactive oxygen species. The body maintains these precise local pH levels through various tiny, built-in systems, like specialized protein pumps that move acids or bases around, and local buffering molecules that soak up excess acid or base. These microenvironments ensure that glutathione and all other cellular components can do their jobs efficiently, even though the overall blood pH stays very stable.

Hydration plays a pivotal role in combating oxidative stress, affecting detoxification, antioxidant production, cellular integrity, and overall health. Coupled with the importance of maintaining electrolytes and proper pH levels, adequate hydration can enhance the body's resilience against oxidative damage. As science continues to illuminate the complexities of hydration and oxidative stress, prioritizing water and electrolyte balance remains a foundational aspect of health and well-being. By recognizing the importance of hydration, we can better support our bodies in achieving optimal function and resilience against the challenges posed by oxidative stress.

Minimizing Exposure to Toxins

In our increasingly industrialized world, exposure to environmental toxins has become a common reality, leading health professionals to stress the importance of minimizing these exposures. Oxidative stress can be exacerbated by toxins present in our personal care products, household items, and yard care supplies. By making conscious choices about the products we use daily, we can significantly reduce our exposure to harmful chemicals and enhance our overall health and well-being.

Personal care products, including cosmetics, skincare items, and fragrances, often contain numerous chemicals that can contribute to oxidative stress. Some common toxins to be wary of include:

- *Parabens*. These are preservatives used in many cosmetics and personal care products (e.g., lotions, shampoos, conditioners). They have been linked to hormone disruption, which can contribute to oxidative stress.[131] Disruption of endocrine system balance (homeostasis) can lead to various disruptions of body system functions as well.

- *Phthalates*. Found in fragrances, nail polish, hair sprays, and many skincare products, phthalates are associated with endocrine disruption and may increase oxidative stress in cells.[132]

- *Formaldehyde-releasing Preservatives.* Used in some hair straightening treatments, shampoos, and nail products, these toxic substances can lead to increased ROS production and have been classified as a carcinogen.[133] Examples include quaternium-15, DMDM hydantoin, diazolidinyl urea, sodium hydroxymethylglycinate, bronopol (2-bromo-2-nitropropane-1,3-diol), polyoxymethylene urea, glyoxal, and imidazolidinyl urea.

- *Triclosan.* Associated with hormone disruption, oxidative stress, and DNA damage, triclosan is found in toothpaste, mouthwashes, deodorants, and cosmetics.[134]

- *Heavy Metals.* Heavy metals like lead, arsenic, and mercury can be found in lipsticks and other color cosmetics. One main reason they are dangerous is that they can create harmful molecules in our body (ROS and RNS).[135] Additionally, these heavy metals have a strong attraction to certain important parts of our body, called thiol groups, which are found in proteins and enzymes that normally help protect us from harm. When heavy metals bind to these protective parts, it disrupts the body's natural defense system. They are linked to neurological damage, kidney and lung toxicity, and other serious health problems.

- *Coal Tar Dyes.* Used as a coloring agent in hair dyes, coal tar dyes are associated with cancer and skin irritation.[136] They are particularly concerning for people with the inherited red cell blood disorder glucose-6-phosphate dehydrogenase deficiency (G6PD deficiency), because they can cause oxidative stress leading to red blood cell breakdown.

- *1,4-Dioxane.* Given the evidence linking 1,4-dioxane to oxidative DNA damage, redox dysregulation, and cancer,[137] it's surprising that this harmful chemical is still used to make products sudsier (e.g., shampoos, liquid soaps).

- *Oxybenzone.* A chemical sunscreen, oxybenzone disrupts the endocrine system and is linked to allergies.[138]

- *Fragrance (Parfum).* Many personal care products are fragranced with synthetic chemicals. Fragrance is a catch-all term that can hide many harmful chemicals, including phthalates, and can cause allergies, skin irritation, and respiratory issues.[139]

To limit exposure, read the labels of personal care products carefully and seek out those made from natural or organic ingredients. Reading labels and opting for products that avoid controversial compounds is a proactive step toward reducing the chemical burden on our bodies.

Household cleaning products can also be significant sources of toxins. Many conventional cleaning agents include harsh chemicals that can irritate the respiratory system and contribute to oxidative stress. Key ingredients to avoid include:

- *Ammonia.* Commonly found in glass cleaners and general-purpose cleaners, ammonia can produce harmful fumes and exacerbate oxidative damage in lung tissue. [140]

- *Chlorine Bleach.* Often used in disinfectants and laundry products, chlorine bleach releases toxic compounds that can lead to increased oxidative stress when inhaled or absorbed by the skin. [141]

- *Fragrances.* Many household products use synthetic fragrances that can contain endocrine disruptors and allergens, contributing to oxidative stress and other health issues.

- *Volatile Organic Compounds (VOCs).* Emitted from a variety of products, including paints, varnishes, cleaning supplies, building materials, and air fresheners, VOCs can trigger the production of free radicals, thereby promoting oxidative stress. [142]

- *Sodium Hypochlorite.* Used as a disinfectant and the active ingredient in chlorine bleach, this chemical has been shown to induce oxidative stress. [143] While pure chlorine may not be well-absorbed through the skin, the disinfection byproducts (DBPs), such as trihalomethanes (THMs) and haloacetonitriles, which form when chlorine reacts with organic matter in water, are known to be absorbed through the skin. Some studies suggest that dermal exposure to these compounds during showering and bathing can be significant, potentially even greater than from drinking water.

Choosing eco-friendly, biodegradable cleaning products is a wise alternative. Many companies now offer non-toxic cleaning solutions made from natural ingredients like vinegar, hypochlorous acid, baking soda, and essential oils that are effective and safe for both the environment and your health.

The products used in home and yard care can also contribute significantly to our cumulative exposure to toxins. Pesticides and herbicides used in lawns and gardens can introduce harmful chemicals that may lead to increased oxidative stress. Here are a few important toxins to avoid:

- *Glyphosate.* A widely used herbicide linked to cancer and other health issues, glyphosate can lead to oxidative stress in soil and plant life, which may also impact human health when residues enter the food chain. [144]

- *Synthetic Fertilizers.* While they promote plant growth, many synthetic fertilizers can disrupt the natural ecosystem in your garden and contribute to increased oxidative stress due to nutrient imbalances in both soil and plants.

- *Petroleum-Based Products.* Used in many pesticides and herbicides, these compounds can persist in the environment and lead to increased ROS production and oxidative stress in exposed organisms.

To reduce exposure to these toxins, consider adopting organic gardening practices. Using compost, natural fertilizers, and integrated pest management techniques can enhance your garden's health while reducing reliance on harmful chemicals. Additionally, native plants can create resilient ecosystems that require fewer chemical inputs.

Minimizing exposure to environmental toxins is essential in combating oxidative stress, a significant contributor to various chronic health conditions. By making mindful choices in personal care, household, and yard products, individuals can reduce their toxic load and promote cellular health. The proactive steps taken today can lead to long-term benefits for personal well-being and the health of future generations. By being vigilant about the chemicals we encounter in our daily lives, we can help our bodies thrive in an increasingly challenging toxic environment.

Manage Stress and Negative Emotions

Recent research has uncovered a vital link between psychological stress, negative emotions, and oxidative stress. Increasingly, stress management practices—such as mindfulness, nature walks, meditation, and prayer—are being recognized for their therapeutic potential in mitigating oxidative stress and promoting overall well-being.

Chronic psychological stress has been shown to elevate levels of ROS in the body.[145] This increase in ROS can outpace the body's natural antioxidants, resulting in oxidative damage to DNA, proteins, and lipids. Moreover, chronic stress can lead to systemic inflammation, which can further amplify oxidative stress. Consequently, individuals under persistent psychological stress may find themselves trapped in a vicious cycle of deteriorating mental and physical health.

Negative emotions, such as anger, anxiety, sadness, and fear, can also contribute to oxidative stress.[146] Although these feelings are a natural part of the human experience, problems arise when they begin to dominate our emotional lives. Emotional distress can provoke physiological responses similar to stress, including increased heart rate and elevated blood pressure, both of which can lead to higher levels of oxidative damage over time. Essentially, excessive negative emotions can maladaptively amplify the fight-

or-flight response, a system designed to protect against real danger, causing it to remain overly sensitive or active even in the absence of a true threat. It means your brain is constantly on high alert, causing you to feel fear or anxiety more often and more intensely than is helpful, leading to distress and problems in daily life. Understanding the interplay between psychological stress, negative emotions, and oxidative stress is essential for addressing both mental and physical health holistically.

Given the significant link between psychological stress, negative emotions, and oxidative stress, incorporating effective stress management practices into daily life can be a powerful means of reducing oxidative damage and promoting well-being.

Mindfulness entails being fully present in the moment, acknowledging thoughts and feelings without judgment. Research has demonstrated that mindfulness can reduce stress, leading to lower levels of cortisol and inflammation. Mindfulness practices, including meditation, have been shown to enhance the body's antioxidant capacity, helping to neutralize excess ROS. [147]

Mindfulness-based stress reduction (MBSR) programs can lead to significant reductions in perceived stress and negative emotions. Regular meditation practice promotes relaxation and emotional regulation, fostering resilient responses to stressors. Studies have illustrated that individuals who engage in mindfulness practices often exhibit decreased levels of oxidative stress markers, such as malondialdehyde (MDA) and increased antioxidant enzyme activity. [148, 149] This research strongly suggests that you must work on your mental and emotional health and manage stress to realize greater wellness.

Nature Walks and Earthing

Connecting with nature through walks in parks, forests, or natural reserves is an impactful way to alleviate stress. Natural settings have been shown to reduce levels of cortisol, promote feelings of tranquility, and enhance overall mood. Nature walks, particularly those in environments rich in biodiversity, can diminish psychological stress, leading to lower oxidative stress levels.

Earthing—the practice of walking barefoot on natural surfaces—has gained popularity for its purported health benefits, including stress reduction and improved well-being. The Earth has a subtle electrical charge, and direct contact with the ground can help stabilize the body's electrical potential. Some studies suggest that earthing may reduce inflammation and oxidative stress markers by promoting a healthier physiological stress response. [150] This is because the surface of the Earth has a virtually unlimited storehouse of accessible electrons, thus it can serve as an electron donor and provide antioxidant benefits. [151] The research shows that grounding an organism to the Earth produces measurable differences in inflammatory markers, such as white blood cells, cytokines and proinflammatory molecules.

Another practice that connects us with the Earth through the respiratory system is forest bathing. Forest bathing, a practice rooted in Japanese tradition (Shinrin-yoku), involves spending time in wooded areas to immerse oneself in the sights, sounds, and smells of the forest. Specifically, researchers conclude that the benefits of this practice are related to the phytoncides (terpenes) emitted by trees, suggesting you can duplicate this practice by diffusing essential oils, which are full of terpenes, in your home or office. Research has shown that forest bathing can lower cortisol levels, improve mood, and enhance overall psychological well-being.[152] The calming effects of being surrounded by nature may help mitigate oxidative damage by reducing stress and promoting relaxation.

Meditation—including practices like guided imagery, transcendental meditation, or loving-kindness meditation—acts as a powerful tool for stress reduction. Such practices encourage emotional resilience and self-awareness, contributing to better regulation of stress responses. Numerous studies have documented the physical and psychological benefits of meditation, including reduced oxidative stress levels.[153]

For instance, individuals who engage in regular meditation often demonstrate lower concentrations of inflammatory cytokines and reduced markers of oxidative stress. This practice fosters a sense of peace that can buffer the impact of external stressors, enabling individuals to respond to life's challenges more effectively.

For many, prayer and spiritual practices serve as a source of comfort and resilience during times of stress. Engaging in prayer, whether as an individual practice or within a community, can provide emotional support and foster a sense of connection and purpose. These aspects can significantly influence mental health and alleviate stress.

Studies exploring the health benefits of prayer and spirituality have suggested that these practices may be linked to lower markers of oxidative stress.[154] By enhancing hope, optimism, and a sense of control, prayer can positively influence an individual's capacity to cope with stress. Moreover, the communal aspect of spiritual practices can provide essential social support and connection, further buffering against stress's deleterious effects.

The interplay between psychological stress, negative emotions, and oxidative stress is complex and profound. Chronic stress can significantly exacerbate oxidative damage and lead to numerous health issues. However, effective stress management practices—such as mindfulness, meditation, nature walks, and prayer—can serve as powerful tools to mitigate oxidative stress. By actively integrating these practices into daily life, individuals can cultivate resilience, enhance their emotional well-being, and foster a more balanced physiological environment. Addressing both the psychological and physical aspects of health offers a holistic approach to enhancing overall well-being and reducing the burden of oxidative stress, paving the way for a healthier future.

Substance use, including smoking, excessive alcohol consumption, and the use of recreational drugs, is intricately linked to increased oxidative stress within the body. This connection is primarily driven by the generation of ROS during the metabolism of these substances and the subsequent depletion of the body's antioxidant defenses.

Tobacco smoke is a well-known source of free radicals that can cause oxidative damage to cellular structures. When tobacco is burned, it releases numerous toxic compounds, including nicotine, tar, carbon monoxide, and a host of other harmful chemicals. The inhalation of cigarette smoke increases the production of ROS in the lungs and throughout the body. [155]

The oxidative stress induced by smoking is significantly higher than that experienced by non-smokers. Studies have shown that smokers exhibit increased levels of oxidative stress markers, such as lipid peroxidation products and depleted antioxidant levels, especially vitamins C and E. Chronic exposure to the oxidative stress from smoking not only leads to respiratory diseases, such as chronic obstructive pulmonary disease (COPD) and lung cancer, but also increases the risk of cardiovascular diseases, neurodegenerative disorders, and premature aging.

Excessive alcohol consumption also poses a significant threat to oxidative balance. [156] Alcohol is metabolized primarily in the liver, where the enzyme alcohol dehydrogenase converts it to acetaldehyde, a toxic intermediate. This metabolic process generates ROS, which can result in direct cellular damage. Additionally, excessive alcohol intake leads to the depletion of glutathione, [157] one of the body's primary intracellular antioxidants. With compromised antioxidant defenses, the body becomes more susceptible to oxidative damage.

Chronic alcohol abuse is linked to a variety of health issues, including liver diseases (such as alcoholic liver disease and cirrhosis), cardiovascular problems, and certain types of cancer. [158] Moreover, alcohol's effects on oxidative stress are compounded by poor dietary choices often associated with excessive drinking, leading to further nutritional deficiencies that can worsen oxidative damage.

Beyond the physical health implications, alcohol can impair judgment, leading to poor decision-making and an increased likelihood of accidents and injuries. It can also contribute to mental health issues, such as depression and anxiety, and can be a factor in social problems, including addiction and domestic violence. Alcohol offers no inherent health benefits—the benefits associated with wine are actually due to compounds in the grapes, not the alcohol itself—and presents a multitude of risks. Therefore, given its potential for harm and lack of positive attributes, avoiding alcohol consumption is the most prudent choice.

The impact of recreational drugs on oxidative stress varies by substance but often leads to increased ROS production and decreased antioxidant activity. For example, stimulant drugs such as cocaine and methamphetamine can elevate oxidative stress levels through mechanisms involving increased heart rate and energy expenditure, which heightens metabolic activity and ROS production.[159, 160] Amphetamine-based stimulant drugs used to treat ADHD and methamphetamine share a core chemical structure of amphetamine. The only difference is an additional "methyl" group (a carbon atom bonded to three hydrogen atoms) in methamphetamine's chemical structure, which dramatically increases its ability to cross the blood-brain barrier. Based on their similar structures, researchers have begun to investigate whether amphetamine-based ADHD drugs will similarly trigger oxidative stress. Although human studies are incomplete, long-term use of stimulant ADHD medications may also pose neurotoxicity risk due to dopamine damage and oxidative stress.[161, 162]

In addition, opioids, while primarily associated with neurobiological effects, may also influence oxidative balance.[163] Chronic use can lead to increased oxidative stress, contributing to various negative health outcomes.

Substance use significantly contributes to oxidative stress through the direct generation of free radicals and the depletion of antioxidants. Smokers, heavy drinkers, and users of recreational drugs are likely to experience higher levels of oxidative stress than their non-user counterparts. Understanding the relationship between substance use and oxidative stress highlights the importance of making informed lifestyle choices. By reducing or eliminating substance use, individuals can not only decrease their exposed oxidative stress levels but also promote better overall health and longevity.

Altogether, this chapter underscores the vital role of lifestyle factors in managing oxidative stress, illustrating that the same habits that support overall health also enhance the body's resilience to oxidative damage. By prioritizing a balanced diet rich in both animal and plant foods, engaging in regular and appropriate physical activity, ensuring adequate sleep, maintaining proper hydration, minimizing toxin exposure, managing stress, and being mindful about substance use, individuals can significantly mitigate oxidative stress and its associated health risks. Each lifestyle choice is interconnected and contributes to a well-rounded approach to health. It's time we move away from seeking quick fixes to address the consequences of our lifestyle choices, and instead fix our lifestyle. As we strive for balance in our day-to-day lives, it becomes clear that nurturing our bodies and minds is essential for promoting longevity, vitality, and overall well-being.

CHAPTER FIVE

Evaluating Antioxidant Potential and Dietary Antioxidants

The consumption of various antioxidants not only helps combat oxidative stress, but is important for maintaining overall health. Each group of antioxidants—vitamins, carotenoids, thiols, enzymatic antioxidants, coenzymes, minerals, bioflavonoids, and hormones—whether naturally produced by the body or consumed, have unique properties and functions that contribute to a comprehensive antioxidant defense system. This diversity not only allows for the neutralization of a wide range of free radicals but also facilitates synergistic interactions that enhance the effectiveness of individual antioxidants. Therefore, a balanced diet with a variety of foods and diversity of supplements is vital for harnessing the full spectrum of antioxidant benefits, promoting optimal health and resilience against chronic diseases.

The use of varying assays to determine antioxidant activity is crucial in the scientific community as it provides a comprehensive understanding of how different compounds interact with free radicals and oxidative stress. Various assays, such as the DPPH, ABTS, and FRAP tests, measure antioxidant capacity through distinct mechanisms, allowing researchers to evaluate the efficacy of specific antioxidants in different conditions and biological systems. This diversity in testing methods enables scientists to identify and quantify the synergistic effects of antioxidants in foods, supplements, and pharmacological agents, guiding research toward effective applications in health promotion and disease prevention. Furthermore, these assays aid in the standardization of antioxidant-rich products for clinical use, ensuring that effective dosages are determined for potential therapeutic interventions, ultimately enhancing the relevance of antioxidants in nutrition and medicine.

For this reason, antioxidant activity is evaluated using several established assays, each offering insight into different mechanisms and applications.

Assays Based on Electron Transfer

DPPH (2,2-diphenyl-1-picrylhydrazyl) Assay. The DPPH assay is a widely used method for evaluating the antioxidant activity of compounds based on their ability to scavenge the stable DPPH free radical. These free radicals are often used in laboratory settings to simulate the behavior of ROS and other free radicals that can be generated in biological systems. DPPH has a distinctive purple color in its stable radical form. When antioxidants react with DPPH, they donate electrons or hydrogen atoms, resulting in a reduction of the DPPH radical and a corresponding change in color, which shifts from purple to a yellow-brown hue. The extent of this color change is quantitatively measured, often using spectrophotometry, a technique that measures how much light a sample absorbs to determine how strong the color change is and indicates the antioxidant capacity of the tested sample. While the DPPH assay primarily assesses the ability of antioxidants to scavenge stable free radicals, it is important to note that it specifically targets free radicals containing nitrogen, as well as other radical species, depending on the antioxidant being tested.

ABTS/TEAC (2,2'-azino-bis(3-ethylbenzothiazoline-6-sulfonic acid)/Trolox Equivalent Antioxidant Capacity) Assay. The ABTS/TEAC assay measures the antioxidant capacity of compounds by generating the ABTS radical cation, which exhibits a distinctive blue-green color. Like DPPH radicals, ABTS/TEAC radicals simulate the effects of ROS in humans. When antioxidants are introduced, they reduce the ABTS radical cation, resulting in a decrease in color intensity. This color change is quantitatively measured, usually by spectrophotometry, and the results are often expressed in terms of Trolox equivalents. Trolox is a water-soluble vitamin E analog that serves as a standard reference compound for comparing antioxidant activity. This assay is particularly versatile because it can assess the ability of antioxidants to scavenge a broad spectrum of free radicals, including those that are soluble in both aqueous and organic solvents.

FRAP (Ferric Reducing Antioxidant Power) Assay. The FRAP assay evaluates the antioxidant capacity of compounds by measuring their ability to reduce ferric ions (Fe^{3+}) to ferrous ions (Fe^{2+}). This reduction process is accompanied by a color change in the reaction mixture, typically shifting from pale yellow to blue or green, depending on the specific conditions of the assay. The intensity of the resulting color change is directly proportional to the antioxidant capacity of the tested sample. Ferric ions (Fe^{3+}) are the oxidized form of iron, which is less reactive compared to ferrous ions (Fe^{2+}), the reduced form. The FRAP assay can simulate important aspects of oxidative stress and ion interactions relevant to human health, by assessing an antioxidant's potential to interact with metal ions in the body and neutralize oxidative threats. In the context of the FRAP assay, antioxidants that can effectively reduce Fe^{3+} to Fe^{2+} demonstrate their ability to counteract oxidative stress.

Assays Based on Hydrogen Atom Transfer

ORAC (Oxygen Radical Absorbance Capacity) Assay. The ORAC assay quantifies the antioxidant capacity of compounds by assessing their ability to protect a fluorescent probe from oxidative damage caused by peroxyl radicals. Peroxyl radicals are a type of ROS that can form during the oxidation of organic compounds, particularly lipids and other biomolecules. Humans are exposed to these radicals via lipid metabolism (creation, breakdown, and transport of fats), the metabolism of chemicals (e.g., herbicides, pharmaceuticals), environmental pollutants (e.g., ozone, particulate matter), cigarette smoke, and even through the consumption of polyunsaturated fatty acids that are low in antioxidants—like some oils, nuts, and processed foods. In this assay, the decay of fluorescence from the probe is monitored over time as it encounters these radicals. The area under the fluorescence decay curve is calculated to determine the ORAC value, which reflects the antioxidant's protective effectiveness. The ORAC assay primarily targets peroxyl radicals, which are significant contributors to lipid peroxidation—a process that can lead to cellular damage and various diseases. By measuring how well antioxidants can inhibit peroxyl radical-induced decay of fluorescence, the ORAC assay provides valuable insights into their potential protective effects against oxidative stress.

HORAC (Hydroxyl Radical Antioxidant Capacity) Assay. The HORAC assay is designed to evaluate the antioxidant capacity of compounds by specifically measuring their ability to neutralize hydroxyl radicals. Humans can be exposed to hydroxyl radicals through several mechanisms, both from natural sources (e.g., UV radiation), biological processes (e.g., byproducts of metabolic processes, inflammatory responses), and human activities (e.g., air pollution, water contamination, smoking). Hydroxyl radicals are highly reactive species that can cause significant oxidative damage to cellular components, including lipids, proteins, and DNA. Unlike the ORAC assay, which assesses protection against peroxyl radicals, the HORAC assay focuses exclusively on the reactivity of antioxidants with hydroxyl radicals—one of the most reactive and damaging free radicals. In this assay, the decrease in fluorescence of a probe, which is sensitive to hydroxyl radical-induced damage, is monitored over time. The extent of fluorescence decay is used to calculate the HORAC value, which reflects the effectiveness of the tested antioxidants in scavenging hydroxyl radicals. The

HORAC assay provides important insights into the protective capabilities of antioxidants against oxidative stress, particularly in contexts where hydroxyl radicals are involved in cellular damage.

β-carotene Bleaching Assay. The β-carotene bleaching assay is designed to assess the ability of antioxidants to inhibit the oxidation of β-carotene, a carotenoid pigment known for its distinctive orange color. In this assay, β-carotene undergoes a color change as it is oxidized, leading to a noticeable fading of its intense color. Antioxidants present in the sample can slow down this oxidation process, thereby preserving the color of β-carotene. The assay primarily targets lipid peroxyl radicals, which are known to initiate lipid oxidation and contribute to the degradation of β-carotene. By measuring the rate of color loss over time, researchers can quantify the antioxidant capacity of the tested compounds, providing insights into their potential protective effects against oxidative damage.

Other Assays

TBARS (Thiobarbituric Acid Reactive Substances) Assay. The TBARS assay is a widely used method for measuring lipid peroxidation by quantifying malondialdehyde (MDA), a key byproduct formed during the oxidative degradation of polyunsaturated fatty acids. The assay is based on the reaction of MDA with thiobarbituric acid, resulting in the formation of a colored complex that can be measured spectrophotometrically. By assessing the concentration of MDA, the TBARS assay provides a reliable indication of lipid peroxidation levels in biological samples. It primarily targets products formed during lipid peroxidation, allowing researchers to evaluate oxidative stress and the potential damage to cellular membranes.

Cellular Antioxidant Activity (CAA) Assay. The CAA assay evaluates the antioxidant activity of compounds within living cells, providing a more physiologically relevant measurement compared to traditional *in vitro* assays. This assay assesses the ability of antioxidants to neutralize intracellular free radicals, which are reactive species that can cause oxidative damage to cellular components such as lipids, proteins, and DNA. By measuring the protective effects of antioxidants against free radical-induced damage in a cellular context, the CAA assay allows researchers to gain insights into the bioavailability and effectiveness of antioxidants in biological systems.

Superoxide Dismutase (SOD) Activity Assay. The SOD activity assay quantifies the enzymatic activity of SOD, an important antioxidant enzyme that catalyzes the dismutation of superoxide radicals. Dismutation is a process where two superoxide radicals (a type of harmful molecule) react together to form a regular oxygen molecule and hydrogen peroxide, effectively reducing their harmful effects in the body. This reaction is crucial for mitigating oxidative stress within cells, as superoxide radicals can cause significant cellular damage if not regulated. By measuring the activity of SOD, this assay provides insights into the cellular defense mechanisms against oxidative damage, specifically targeting the scavenging of superoxide radicals.

Catalase (CAT) Activity Assay. The CAT activity assay measures the enzymatic activity of catalase, an enzyme that catalyzes the decomposition of hydrogen peroxide into water and oxygen. This reaction is vital for cellular protection against oxidative stress, as hydrogen peroxide is a potentially harmful reactive oxygen species that can damage cellular components. By quantifying catalase activity, this assay provides insights into the capacity of cells to mitigate oxidative damage by reducing hydrogen peroxide levels, thereby contributing to the overall antioxidant defense system.

Total Antioxidant Capacity (TAC) Assay. The TAC assay measures the cumulative ability of antioxidants present in a sample to scavenge free radicals or neutralize other reactive species. Unlike specific assays that target individual radicals or components (such as DPPH or ABTS), the TAC assay provides a comprehensive assessment of the overall antioxidant potential of a mixture, making it particularly useful for evaluating complex biological samples, such as food extracts, biological fluids, or plant materials. The TAC assay is valuable because it provides an overall measurement of antioxidant activity, especially important in naturally occurring mixtures where different antioxidant compounds may work synergistically. TAC values also allow researchers and clinicians to compare antioxidant capacities of various foods, supplements, or biological samples, highlighting those with higher protective potential against oxidative stress.

Dietary Antioxidants

As discussed earlier, oxidative stress plays a crucial role in the pathogenesis of various chronic diseases, including cardiovascular diseases, neurodegenerative disorders, and cancer. Understanding the nutritional aspects of antioxidant defense is vital for developing strategies to mitigate oxidative damage.

In addition to making its own antioxidants (endogenous), the body relies on a consistent supply of antioxidants from the diet or supplementation (exogenous). While our bodies have robust endogenous antioxidant systems, several factors can lead to an imbalance between free radical production and antioxidant defenses, making exogenous antioxidants beneficial. Increased free radical production stemming from environmental pollutants, UV radiation, lifestyle stressors like poor diet and lack of exercise, and even the natural process of aging can overwhelm the body's defenses. Additionally, the body's own antioxidant reserves, such as glutathione and CoQ10, can become depleted under conditions of heightened oxidative stress, particularly when coupled with nutrient deficiencies that impair their production. Furthermore, endogenous systems, while effective, may not be sufficient to neutralize all free radicals, especially during periods of intense physical activity or illness, where free radical generation is temporarily elevated. Exogenous antioxidants provide vital supplementary support, working synergistically with endogenous antioxidants to bolster overall protection. Certain exogenous antioxidants also offer targeted benefits to specific tissues or organs and can work in ways the body's own antioxidants do not. Essentially, incorporating exogenous antioxidants into a healthy lifestyle helps maintain a critical balance, safeguarding against the detrimental effects of oxidative stress and promoting overall well-being.

When it comes to protecting our bodies from oxidative stress, one of the biggest keys to effective antioxidant consumption—whether through diet or supplementation—is obtaining a variety of antioxidants. Different antioxidants have different mechanisms of action, target specific free radicals, and complement each other's effects. This diversity is crucial for forming a robust defense against oxidative damage and ensuring that our bodies can respond effectively to the multitude of free radicals generated by metabolic processes and environmental factors. Below, the various groups of antioxidants will be explored, detailing their differences, types, and functions.

Our antioxidant defense system is made up of those naturally produced by the body (endogenous) and those we get from dietary sources (food, beverages, and supplements). Antioxidants obtained directly from foods we eat play a crucial role in neutralizing free radicals and preventing oxidative damage.

Vitamin C, or ascorbic acid, is a water-soluble antioxidant known for its ability to neutralize a wide range of free radicals, particularly hydroxyl radicals and singlet oxygen. It is vital for collagen synthesis, immune function, and promoting the absorption of non-heme iron. One of its key roles is to regenerate vitamin E, which enhances the overall antioxidant defense system. When vitamin E neutralizes free radicals, it is converted to a less active form. Vitamin C can restore vitamin E to its active state, thereby maintaining its antioxidant potential. Through these mechanisms, vitamin C protects cells from oxidative damage while supporting various physiological processes.

The vitamin E complex consists primarily of tocopherols (alpha-, beta-, gamma-, and delta-tocopherol) and tocotrienols (alpha-, beta-, gamma-, and delta-tocotrienol), the most active being alpha-tocopherol. This fat-soluble vitamin primarily protects cell membranes from lipid peroxidation by quenching lipid radicals. Upon encountering free radicals, vitamin E donates an electron to stabilize the radical, thus preventing damage to cellular membranes. Vitamin E is crucial for maintaining the integrity of cellular structures and has immune-supporting properties. Like vitamin C, it operates best in a synergistic manner; by working alongside vitamin C and other antioxidants, vitamin E's longevity and protective effects are enhanced, preventing oxidative stress in lipid-rich environments such as cell membranes.

Carotenoids are a class of pigments found in colorful fruits and vegetables. These fat-soluble antioxidants—which include beta-carotene, lutein, and zeaxanthin—act to scavenge free radicals such as singlet oxygen and can convert to vitamin A, serving vital roles in vision and skin health. Beyond their antioxidant properties, carotenoids contribute to immune function and cardiovascular health. Their ability to protect against oxidative damage is especially important for the eyes, where lutein and zeaxanthin are concentrated and play protective roles against light-induced damage. Beta-carotene and vitamin E can work synergistically. While beta-carotene neutralizes free radicals, it can also help regenerate the oxidized form of vitamin E, enhancing the overall antioxidant effect.

Bioflavonoids, a diverse class of plant compounds, exhibit powerful antioxidant properties. Common examples include quercetin, catechins, and anthocyanins. These compounds can scavenge free radicals, chelate metals, and modulate cellular signaling pathways, helping reduce inflammation and support immune function. By reinforcing the antioxidant capacity of vitamins and other nutrients, bioflavonoids help maintain overall health and protect against chronic diseases.

Quercetin, along with other major bioflavonoids such as kaempferol, myricetin, and hesperidin, offers significant antioxidant benefits that contribute to overall health. These powerful compounds are known for their ability to neutralize free radicals, thereby reducing oxidative stress and minimizing cellular damage. Quercetin, in particular, has been shown to possess anti-inflammatory properties, which can help alleviate chronic inflammation linked to various diseases, including heart disease and certain cancers. Additionally, quercetin and its bioflavonoid counterparts can enhance immune function, improve cardiovascular health by promoting healthy blood pressure and cholesterol levels, and even support respiratory health by mitigating allergy symptoms. The synergistic effects of these bioflavonoids contribute to their potential in preventing chronic diseases and promoting overall well-being, making them valuable components of a healthy diet rich in fruits, vegetables, and herbal sources.

Flavonoids, including quercetin and other bioflavonoids, interact synergistically with vitamin antioxidants, enhancing their effectiveness and bioavailability in several ways. One of the key mechanisms is the ability of flavonoids to increase the bioavailability of vitamin C. Quercetin, for instance, can protect vitamin C from being broken down in the gastrointestinal tract, thereby allowing more of it to be absorbed into the bloodstream. Additionally, flavonoids can regenerate oxidized forms of vitamin C, effectively recycling it back to its active state after it has neutralized free radicals. This recycling process ensures that vitamin C remains available to exert its antioxidant effects for a longer duration. Furthermore, the presence of flavonoids can enhance the absorption of other essential nutrients, improving the overall antioxidant capacity within the body.

Quercetin and other flavonoids can enhance cellular delivery of zinc through several mechanisms. One key mechanism is by acting as a chelator, meaning they can bind to zinc ions and facilitate their transport across cellular membranes. This chelation not only stabilizes zinc in its soluble form but also helps in its absorption and utilization within cells.

Additionally, quercetin and flavonoids can modulate various transport proteins, such as zinc transporters, which regulate the uptake of zinc into cells. For instance, by influencing the expression or activity of transport proteins like ZIP (Zrt-, Irt-like protein) or ZnT (Zinc Transporter), flavonoids can promote an increase in zinc uptake, thus enhancing cellular zinc levels.

Furthermore, the antioxidant properties of quercetin may help protect zinc in its active form from being oxidized, ensuring its bioavailability for critical biochemical processes. This synergistic relationship is particularly important since zinc plays a vital role in numerous cellular functions, including enzymatic reactions and immune responses. The combination of quercetin or other flavonoids with zinc can thus optimize its cellular delivery and function, highlighting the importance of incorporating a variety of nutrients in the diet.

Catechins are a type of flavonoid found primarily in tea, especially green tea, as well as in foods like cacao, apples, and berries, and they offer significant antioxidant benefits. One of their key roles is to neutralize free radicals that can lead to oxidative stress and cellular damage. By reducing oxidative stress, catechins also exhibit anti-inflammatory effects, potentially lowering the risk of chronic diseases. They are linked to cardiovascular health through improvements in endothelial function, reductions in blood pressure, and decreased oxidation of LDL cholesterol. Furthermore, catechins may have cancer-preventive properties, as some studies indicate they can inhibit the growth of cancer cells and promote programmed cell death in certain cancers.

Additionally, they may enhance fat oxidation and improve insulin sensitivity, supporting weight management and metabolic health. Overall, the antioxidant properties of catechins play a significant role in promoting overall health and reducing disease risk.

Catechins have been shown to interact with other antioxidants in ways that involve recycling and enhancing cellular delivery. One of the key mechanisms through which catechins operate is the regeneration of vitamin C, where they can help convert oxidized forms of vitamin C (ascorbic acid) back to its reduced form. This recycling effect not only enhances the availability of vitamin C, a potent antioxidant on its own, but also maintains higher levels of antioxidant activity within cells. Additionally, catechins may interact with other antioxidants, such as vitamin E, protecting it from oxidation and thus preserving its effectiveness, further contributing to the overall antioxidant defense system in the body.

Catechins can also enhance the delivery of certain minerals, like zinc, into cells by modulating the activity of specific transport proteins. While the detailed mechanisms are still under investigation, the chelating properties of catechins may facilitate the bioavailability and cellular uptake of important nutrients. Furthermore, catechins can impact the expression and activity of transporters responsible for nutrient uptake, potentially leading to increased delivery of antioxidants and other essential compounds into cells. By mitigating oxidative stress, catechins may improve cellular health, optimizing the uptake of various nutrients through enhanced cell membrane fluidity and function. Overall, these mechanisms highlight the role of catechins and other flavonoids in contributing to the antioxidant network in the body, reinforcing the benefits of a diet rich in these compounds and their potential to support cellular health and function.

Anthocyanins belong to the flavonoid class of compounds, which are a specific subclass of polyphenols characterized by their vibrant red, purple, and blue pigments found in various fruits, vegetables, and flowers. Commonly recognized for their antioxidant properties, anthocyanins play a key role in neutralizing free radicals and mitigating oxidative stress. These compounds are abundant in foods such as blueberries, blackberries, red cabbage, and cherries. Beyond their antioxidative capabilities, anthocyanins possess anti-inflammatory effects and have been linked to various health benefits, including improved cardiovascular health, enhanced cognitive function, and potential protective effects against certain chronic diseases. They exert these beneficial effects through mechanisms such as modulating inflammatory pathways, promoting endothelial function, and improving cholesterol metabolism. The growing body of research on anthocyanins highlights their importance not only as dietary components but also as potential therapeutic agents for maintaining health and preventing disease.

While bioflavonoids were covered, it is also relevant to note that polyphenols encompass a broader category beyond flavonoids. This group includes compounds such as resveratrol, curcumin, and ellagic acid, which exhibit various antioxidant properties. Polyphenols play a significant role in combating oxidative stress, reducing inflammation, and supporting cardiovascular health through multiple mechanisms, including scavenging free radicals and modulating cell signaling pathways.

One of the remarkable aspects of polyphenols, including both flavonoids and non-flavonoid compounds, is their ability to interact synergistically with other antioxidants in the body. For instance, anthocyanins can participate in the recycling of other antioxidants, such as vitamin C. When vitamin C undergoes oxidation, anthocyanins can help restore it to its active form, thus enhancing its antioxidant capacity and prolonging its effectiveness within cellular environments. This recycling capability ensures that antioxidants remain available for continual defense against oxidative damage.

Moreover, polyphenols can also enhance the absorption and delivery of other important antioxidants, such as vitamin E and certain minerals like selenium and zinc. They can modulate the expression and activity of specific transport proteins responsible for nutrient uptake, effectively increasing the bioavailability of these vital compounds. For instance, certain polyphenols can influence the cellular uptake of vitamin E by protecting it from oxidation, thereby enhancing its capacity to neutralize free radicals.

Additionally, polyphenols are known to influence cellular signaling pathways that regulate antioxidant defenses, such as the Nrf2 pathway, which activates the expression of various antioxidant enzymes. By upregulating these protective pathways, polyphenols not only bolster the body's antioxidant capabilities but also enhance overall resilience against stressors that contribute to chronic inflammation and cardiovascular diseases.

The anti-inflammatory properties of polyphenols, including resveratrol and curcumin, further complement their antioxidant actions. These compounds can inhibit pro-inflammatory signaling pathways and cytokines, reducing the overall inflammatory load on the body. This dual action of scavenging free radicals while simultaneously modulating inflammation is crucial for maintaining overall health and longevity.

Resveratrol is a polyphenolic compound found predominantly in the skin of red grapes, berries, and peanuts. Known for its antioxidant activity, resveratrol combats oxidative stress by neutralizing harmful free radicals in the body. This action helps to protect cells from damage and supports overall health. Studies suggest that resveratrol may also enhance the activity of other antioxidants, contributing to a synergistic effect that amplifies its protective capabilities. Additionally, it has been linked to various health benefits, including cardiovascular protection and anti-inflammatory properties, further emphasizing its role as a powerful dietary antioxidant.

Curcumin is an active compound in turmeric, a spice commonly used in Indian cuisine. It is recognized for its potent antioxidant properties, which enable it to scavenge free radicals and reduce oxidative stress in the body. Curcumin promotes the expression of several antioxidant enzymes, which bolster the body's natural defense mechanisms against oxidative damage. In addition to its antioxidant effects, curcumin possesses anti-inflammatory and anticancer properties, making it a promising candidate for supporting overall health. Its bioavailability can be a limiting factor, but combining it with the essential oil, galactomannan, or piperine can significantly enhance its absorption.

Ellagic acid is a naturally occurring polyphenol found in various fruits and nuts, particularly in strawberries, raspberries, and walnuts. Known for its antioxidant activity, ellagic acid helps to neutralize free radicals, thereby reducing oxidative stress and cellular damage. Research indicates that ellagic acid not only possesses direct antioxidant properties but also enhances the body's enzymatic defenses against oxidative stress by inducing the expression of antioxidant enzymes. Moreover, ellagic acid is associated with various potential health benefits, including anti-inflammatory and anticancer effects, highlighting its significance as a dietary antioxidant that supports overall wellness.

In summary, polyphenols—encompassing a wide variety of compounds including flavonoids—provide a multifaceted approach to combat oxidative stress and promote health. Their interactions with other antioxidants through recycling and enhanced delivery mechanisms, alongside their ability to modulate cellular signaling pathways, underscore their significance in nutritional biochemistry and their potential as therapeutic agents in preventing chronic disease. With ongoing research continuing to unveil the complexities of these interactions, the inclusion of polyphenol-rich foods in the diet remains a key recommendation for anyone seeking to enhance their health through natural means.

Alpha-lipoic acid (ALA) is considered a dietary antioxidant, as it is a naturally occurring compound that plays a crucial role in metabolism and possesses antioxidant properties that contribute to various health benefits. ALA is capable of scavenging free radicals, and by neutralizing them, ALA helps protect against oxidative stress. One of the unique features of alpha-lipoic acid is its solubility in both water and fat, allowing it to function effectively in various cellular environments and making it a versatile antioxidant that combats oxidative stress throughout the body.

Additionally, ALA has the ability to regenerate other antioxidants, such as vitamins C and E, restoring them to their active forms after they have been oxidized.[164,165] This recycling process significantly enhances the overall antioxidant capacity of the body. ALA also plays a vital role in mitochondrial function, where it aids in the production of energy (ATP). Since mitochondria are significant sources of ROS, ALA contributes to controlling oxidative stress within these critical cellular organelles. Studies have shown that ALA may offer various health benefits, such as improving insulin sensitivity, supporting nerve health in individuals with diabetic neuropathy, and reducing inflammation, with its antioxidant properties being a key component of these effects.[166]

Dietary sources of alpha-lipoic acid include red meat, organ meats like liver, as well as spinach, broccoli, tomatoes, potatoes, and yeast. Although ALA is available from food sources, it is also commonly sold as a dietary supplement for those looking to enhance their antioxidant intake. In summary, ALA is an important dietary antioxidant that plays an essential role in protecting cells from oxidative damage, regenerating other antioxidants, and supporting overall metabolic health.

Certain minerals and trace elements, such as selenium, zinc, and manganese, function as essential components of various antioxidant enzymes. Selenium is a key player in glutathione peroxidase, which detoxifies peroxides, while zinc serves as a cofactor for SOD, making them both essential for managing cellular oxidative stress. Manganese is also integral to SOD activity in mitochondria. These minerals support the body's innate antioxidant systems, proving crucial for optimal cellular function and protection against oxidative damage.

They also possess their own antioxidant benefits. Selenium is a trace mineral that plays a vital role in the body's antioxidant defense system. In addition to being a key component of glutathione peroxidase, selenium is also a critical part of several other important antioxidant enzymes, including thyroid peroxidase, selenoproteins P and W, and methionine sulfoxide reductase.[167] By supporting the function of these enzymes, selenium contributes to reducing oxidative stress.

Additionally, adequate selenium levels have been associated with improved immune function and may provide protective benefits against inflammation, underscoring its importance in maintaining overall health.

Zinc is an essential mineral that serves numerous biological functions, including its crucial role as an antioxidant. It contributes to the body's antioxidant defense by stabilizing cell membranes and protecting tissues from oxidative stress. Beyond SOD, zinc is a required cofactor for the function of various antioxidant enzymes, such as alkaline phosphatase, carbonic anhydrase, and thioltransferase. By helping to combat oxidative stress, zinc supports immune function and plays a role in wound healing. Furthermore, adequate zinc intake is linked to a reduced risk of chronic diseases, emphasizing its significance as a key dietary antioxidant for maintaining optimal health.

Manganese is an essential trace element that functions as a cofactor for several antioxidant enzymes, particularly manganese superoxide dismutase (MnSOD), which is vital for protecting mitochondria from oxidative damage. This enzyme helps catalyze the conversion of harmful superoxide radicals into less damaging molecules, thus mitigating oxidative stress within cells. Manganese also contributes to the overall antioxidant defense system by supporting other antioxidant mechanisms in the body. Adequate manganese levels are associated with various health benefits, including anti-inflammatory properties and potential protection against conditions linked to oxidative stress, such as neurodegenerative diseases.

Food sources of antioxidants.

- *Vitamin C*: citrus fruits, berries, peppers, cruciferous vegetables, kiwi, mango, pineapple, cantaloupe, watermelon
- *Vitamin E*: Olive oil, wheat germ, nuts, sunflower seeds, green leafy vegetables, avocado
- *Carotenoids*: Carrots, sweet potatoes, pumpkin, cantaloupe, mango, apricots, orange and red bell peppers, tomatoes, pink grapefruit, dark leafy greens, broccoli, eggs
- *Bioflavonoids*: citrus fruits, grapes, berries, onions, apples, grapes, cocoa, leafy greens, celery, legumes
- *Polyphenols*: Berries, cocoa, apples, plums, cherries, grapes, pomegranates, nuts, olives, onions, spinach, broccoli, spices (turmeric, cloves, cinnamon, ginger, cumin)
- *Alpha-lipoic Acid*: Red meat (especially organ meats), brocooli, tomatoes, Brussels sprouts, peas, carrots, potatoes, yeast, rice bran

- *Selenium*: Brazil nuts, seafood, meats (especially dark meat and organ meats), mushrooms, eggs, dairy products, legumes, whole grains

- *Zinc*: Red meat, shellfish, poultry, oysters, legumes, nuts, seeds, eggs, dairy products, whole grains

- *Manganese*: Whole grains, nuts, legumes, shellfish, leafy greens, pineapple, berries, spices (black pepper, turmeric, cloves), avocado, cocoa

Endogenous Antioxidants

Endogenous antioxidants are crucial to the body's intrinsic defense system against oxidative stress, acting to neutralize harmful free radicals generated during metabolic processes. These antioxidants can be classified into several categories, including enzymatic antioxidants like SOD and CAT, which catalyze biochemical reactions to detoxify reactive oxygen species. Additionally, antioxidant coenzymes such as coenzyme Q10 (CoQ10) and lipoic acid play pivotal roles not only in energy production but also in bolstering the antioxidant network by regenerating other antioxidants. Other important elements include ubiquinol, the reduced form of CoQ10, and thiols like glutathione, recognized for their crucial detoxifying capabilities. Furthermore, antioxidant hormones such as melatonin and estrogen contribute to the body's protective mechanisms against oxidative damage. Together, these endogenous antioxidants form a complex and interrelated system that effectively mitigates oxidative stress, promoting cellular health and overall well-being.

Our body has a way of controlling how many antioxidants it makes and how active these antioxidants are. Think of antioxidants as the body's own little superheroes that fight off harmful substances (free radicals), which can damage our cells. One of the key partners in this process is a protein called Nrf2. When our body is under stress, like when there are too many free radicals around, Nrf2 moves into the part of the cell that acts like a control center, called the nucleus. Once there, Nrf2 switches on certain genes that produce antioxidant enzymes and proteins, helping to fight off the damage. This system is important because it helps keep our cells balanced and healthy, which is essential for a long and healthy life. Understanding how this works highlights why it's important to get both antioxidants from our body's own production and from the foods we eat to stay strong and healthy against damage from free radicals.

Enzymatic antioxidants are proteins that catalyze biochemical reactions to neutralize free radicals. Key examples include SOD and CAT. GPX, another essential enzyme, uses glutathione to detoxify peroxides. These enzymes enhance the body's endogenous antioxidant defense system, effectively targeting and mitigating the harmful effects of reactive oxygen species generated during metabolic processes.

Antioxidant coenzymes, such as CoQ10 and lipoic acid, play critical roles in energy production within cells and act as antioxidants. CoQ10 is involved in mitochondrial energy metabolism and can neutralize free radicals, while lipoic acid is unique in being both water- and fat-soluble, allowing it to trap various types of free radicals. These coenzymes also help regenerate other antioxidants, bolstering the overall antioxidant defense network and reinforcing energy production and cellular health.

Ubiquinol is the reduced form of CoQ10. While CoQ10 was mentioned as an antioxidant coenzyme, ubiquinol is notable for its superior antioxidant capacity, particularly in mitochondrial membranes, where it protects against oxidative damage and promotes energy production.

Thiols, such as glutathione, cysteine, and N-acetylcysteine (NAC), are compounds characterized by their sulfur-containing groups, allowing them to participate in redox reactions. Glutathione, is an endogenous thiol-based antioxidant, often dubbed the "master antioxidant." It is crucial for detoxifying harmful substances and recycling other antioxidants, like vitamins C and E. Thiols help mitigate oxidative stress by donating electrons to neutralize free radicals and maintaining cellular redox balance. NAC and cysteine are obtained through diet or supplementation. Their role in detoxification processes further underscores the importance of thiols in combating oxidative damage.

Antioxidant hormones, such as melatonin and estrogen, also play significant roles in reducing oxidative stress. Melatonin is known for its potent antioxidant properties, effectively neutralizing a variety of free radicals and regulating the endogenous antioxidant system.[168] Estrogen, while primarily recognized for its role in reproductive health, has also been shown to exhibit antioxidant effects, thereby providing additional protective benefits against oxidative stress, particularly in cardiovascular health.

The interplay between our body's endogenous antioxidant system and the intake of exogenous dietary antioxidants is integral in the ongoing fight against oxidative stress and its damaging repercussions on health. While our bodies possess remarkable mechanisms to produce antioxidants like glutathione and superoxide dismutase, they can be overwhelmed by the insidious effects of environmental toxins, poor lifestyle choices, and the natural aging process. This underscores the necessity of complementing endogenous defenses with a robust intake of diverse dietary antioxidants from a variety of fresh, whole foods. The synergistic interactions between these compounds not only enhance their individual efficacy but also fortify our biological systems against oxidative damage. Emphasizing a diet rich in varying antioxidants empowers individuals not only to maximize their health potential but also to cultivate resilience against the chronic diseases often triggered by oxidative stress. As the body relies on both its internal production and external sources of antioxidants, fostering this dynamic through conscientious dietary choices is essential for maintaining optimal health and well-being in a world fraught with oxidative challenges.

CHAPTER SIX

Dietary Supplements and Nutraceuticals as Antioxidants

In addition to obtaining antioxidants from a well-balanced diet, we can also enhance our antioxidant support through dietary supplements. These supplements can provide a concentrated source of various nutrients or compounds that are known for their powerful antioxidant properties. For instance, vitamin C and vitamin E are commonly found in supplement form and can help combat oxidative stress, bolstering the body's defenses against free radicals. Minerals such as selenium and zinc, which play critical roles in enzyme function and antioxidant activity, are also available as supplements, ensuring that individuals meet their recommended intake even if their dietary sources are insufficient.

Furthermore, there are several endogenous antioxidants that the body produces, such as glutathione, which can also be supplemented to enhance the body's overall antioxidant capacity. Another way to support glutathione is to supplement with cofactors like N-acetylcysteine (NAC), which can help boost glutathione levels, promoting detoxification and supporting cellular health. Additionally, compounds like alpha-lipoic acid and coenzyme Q10 are popular supplements that not only serve as antioxidants but also support mitochondrial function and energy production, thus providing a dual benefit.

Dietary supplementation of antioxidants can be particularly beneficial for individuals with dietary restrictions, insufficiencies or deficiencies in key antioxidants, increased antioxidant needs due to environmental exposures, chronic health conditions, or those seeking to enhance athletic performance and recovery. However, it's important to approach supplementation with an awareness of balance and avoid overconsuming large amounts of a single antioxidant. While antioxidants are crucial for neutralizing harmful free radicals, excessive consumption of a single antioxidant can disrupt the delicate balance of physiological processes. Overloading the body with one type can interfere with the natural functions of other antioxidants, potentially creating an imbalance that promotes oxidative stress rather than preventing it. Furthermore, high doses of certain antioxidants have been linked to adverse effects, including an increased risk of certain cancers, interference with medication, and digestive issues. Informed healthcare professionals can tailor supplementation to an individual's specific health needs, dietary habits, and lifestyle factors, ensuring an effective strategy for bolstering antioxidant defenses and supporting overall health.

Below are some of the most potent and valuable antioxidant supplements. The preferred form(s), adult dosing, and potential companion supplements are listed for reference.

Vitamin C

Preferred Form(s): Liposomal vitamin C | *Typical Adult Dosage*: Maintenance—500–1,000 mg/day in divided doses; Intensive—2,000 mg/day in divided doses | *Companion Nutrients*: Bioflavonoids, vitamin E, iron, zinc, vitamin D | *When to Take*: With or without food any time

Vitamin C is a powerful water-soluble antioxidant that primarily combats ROS such as hydroxyl radicals and superoxide anions. In nature, it exists as a complex of compounds with ascorbic acid as the core and bioflavonoids that enhance its absorption and antioxidant effects. Supplements often use mineral ascorbates (e.g., sodium ascorbate, calcium ascorbate), which are less acidic and gentler on the stomach. Some use the fat-soluble form of vitamin C—ascorbyl palmitate, allowing it to integrate intro cell membranes, potentially enhancing tissue retention. To improve the absorption and retention of vitamin C, it should be taken in liposomal form or with lipid metabolites, and with citrus bioflavonoids. Liposomal vitamin C is 1.68–1.77 times more bioavailable than non-liposomal vitamin C,[169] and some manufactures boasting 3 to 12.17 times greater bioavailability.[170,171] It is also retained in the body for approximately 30–50% longer.[172,173,174] This ultimately leads to faster, more pronounced, and sustained beneficial effects. Vitamin C is particularly beneficial in reducing oxidative stress linked to conditions like cardiovascular disease, skin aging, and certain cancers. Additionally, vitamin C supports immune function and aids in the regeneration of other antioxidants, such as vitamin E.

A meta-analysis of randomized clinical trials examined the impact of vitamin C supplementation on oxidative stress following acute exercise.[175] The study found that vitamin C significantly reduced lipid peroxidation and suppressed the inflammatory response, specifically interleukin-6 (IL-6), after exercise. However, it did not affect muscle soreness or strength, indicating a specific role in reducing oxidative stress and inflammation rather than improving muscle recovery.

One study found that liposomal vitamin C is more rapidly absorbed and leads to higher serum concentration of vitamin C, resulting in a greater reduction in both oxidative stress and inflammatory markers.[176] Remarkably, these results were achieved after a single 1000 mg dose of liposomal vitamin C and after maintaining a low vitamin C diet for the previous 14 days.

A pilot randomized clinical trial investigated the effects of vitamin C (500 mg/day), alone and in combination with resveratrol (500 mg/day), on oxidative stress in postmenopausal women.[177] The study reported significant reductions in oxidative stress markers such as lipohydroperoxides and malondialdehyde, along with an increase in total antioxidant capacity, highlighting the potential of vitamin C in managing oxidative stress in postmenopausal women. Interestingly, the vitamin C group experienced better reductions in oxidative stress than the resveratrol and resveratrol plus vitamin C groups.

A study on overweight students demonstrated that vitamin C supplementation, along with walking, reduced oxidative stress markers like ROS and 8-OHdG and inflammatory markers such as IL-1β.[178] Study participants took 100 mg of vitamin C after each meal and walked 6,000 steps every afternoon for two weeks. This suggests that vitamin C can be an effective intervention for reducing oxidative stress and inflammation in overweight individuals when combined with walking.

Clinical research, particularly meta-analyses and systematic reviews, supports the role of vitamin C in reducing oxidative stress, especially in the context of exercise and chronic conditions. While vitamin C alone has shown significant effects in reducing specific oxidative stress markers, its combination with other antioxidants like vitamin E may enhance these effects. These findings underscore the potential of vitamin C as a therapeutic agent in managing oxidative stress-related conditions.

Vitamin E (Tocopherol/Tocotrienol)

Preferred Form(s): Mixed natural tocopherols | *Typical Adult Dosage*: Maintenance—300–500 IU (200–450 mg) daily; Intensive—1000–1500 IU (670–1000 mg) daily | *Companion Nutrients*: Vitamin C, selenium, carotenoids, omega-3 fatty acids | *When to Take*: With food in the morning

Vitamin E consists of a group of fat-soluble compounds known as tocopherols and tocotrienols with antioxidant properties, primarily protecting against lipid peroxidation caused by free radicals, especially peroxyl radicals. You can find it in natural (d-alpha-tocopherol) and synthetic forms (dl-alpha-tocopherol), with the natural form being more bioavailable. The amount of vitamin E you absorb depends on what you eat it with. Alpha-tocopherol is preferred by the body since it is the most active form of vitamin E in the human body. The alpha-tocopherol transfer protein (α-TTP) plays a crucial role in vitamin E bioavailability, specifically concerning alpha-tocopherol. It is primarily located in the liver and preferentially selects alpha-tocopherol from other forms of vitamin E. This process is vital for ensuring that alpha-tocopherol is efficiently incorporated into very low-density lipoproteins, which transport it to other tissues in the body. Optimum functioning of α-TTP is essential for maintaining healthy vitamin E levels in the blood and tissues. If it is not functioning efficiently, the body struggles to distribute vitamin E to the tissues that need it. This can lead to a paradoxical situation where tissues are deficient in vitamin E but blood tests may show normal or even high levels of vitamin E. Gamma-tocopherol is metabolized and excreted faster by the body, while tocotrienols tend to have higher antioxidant activity but lower bioavailability.[179,180] Self-nanoemulsifying forms of mixed tocotrienols can improve bioavailability up to three times.[181] It helps guard against heart disease, age-related cognitive decline, and skin damage due to UV exposure. Vitamin E works best in conjunction with Vitamin C, enhancing the body's overall antioxidant defense system. Despite potential differences in absorption rates, consuming a mixture of tocopherols and tocotrienols provides a broader spectrum of vitamin E's potential health benefits.

A systematic review and meta-analysis evaluated the combined effects of omega-3 fatty acids and vitamin E on oxidative stress parameters.[182] The study found significant improvements in total antioxidant capacity and nitric oxide levels, along with a reduction in malondialdehyde levels. However, no significant changes were observed in endogenous antioxidants (glutathione, superoxide dismutase, and catalase) activities, suggesting that while vitamin E can enhance certain antioxidant parameters, its effects may be limited to specific oxidative stress markers when combined with omega-3.

A meta-analysis investigated the role of vitamin E in reducing exercise-induced muscle damage and oxidative stress.[183] The findings showed that low-dose (< 500 IU) vitamin E supplementation could protect against muscle damage, particularly in athletes, by reducing creatine kinase and lactate dehydrogenase levels, indicating its protective role against oxidative stress from physical exertion. A review of the combined effects of vitamins C and E on oxidative stress and inflammation post-exercise found that this combination diminished lipid peroxidation and inflammatory markers, suggesting a synergistic effect in reducing oxidative damage and promoting recovery.[184] It was effective when taken before exercise.

A study among healthy Asian individuals under age 35 found that vitamin E supplementation improved cell-mediated immunity and reduced oxidative stress markers, indicating its potential benefits in enhancing immune function and reducing oxidative damage. Specifically, oxidative stress in T cells was reduced, which improves their function. T cells primarily identify and destroy harmful pathogens, infected cells, and abnormal cells (like cancer cells), while also regulating other immune responses. T cells are closely associated with oxidative stress, as both low and high levels of ROS can influence their function, differentiation, and survival, playing a complex role in immune responses and various diseases. The participants took 400 IU (233 mg) of vitamin E daily after dinner for 28 days.

A meta-analysis focused on the effects of vitamin E supplementation—with the most common dose being 400 IU daily—in patients undergoing hemodialysis, a group particularly vulnerable to oxidative stress and inflammation.[185] The results indicated that vitamin E significantly reduced levels of intercellular and vascular cell adhesion molecules, C-reactive protein, and malondialdehyde, highlighting its potential to alleviate oxidative stress and both vascular and systemic inflammation in these individuals.

In summary, vitamin E is a vital antioxidant that serves multiple essential functions in the human body, particularly regarding immune support and protection against oxidative stress. Its efficiency is closely linked to its forms, bioavailability, and interactions with other nutrients such as vitamin C and omega-3 fatty acids. The research highlights not only its protective roles in physical activities and among vulnerable populations, such as those undergoing hemodialysis, but also suggests that the interplay between various antioxidants can enhance overall health outcomes. By ensuring adequate intake and understanding the mechanisms that govern its bioavailability, we can better harness the potential of vitamin E in promoting longevity and well-being, ultimately contributing to a healthier and more resilient body.

Alpha-Lipoic Acid (ALA)

Preferred Form(s): Stabilized Alpha-Lipoic Acid (Na-RALA) or liquid R-ALA | *Typical Adult Dosage*: Liquid RALA—50-100 mg; Na-RALA—100-600 mg/day; standard ALA—600 mg, up to three times daily | *Companion Nutrients*: Vitamin C, vitamin E, glutathione, B vitamins, biotin, CoQ10, carnitine, NAC, fish oil | *When to Take*: Morning on an empty stomach

Alpha-lipoic acid (ALA) is a distinctive and versatile antioxidant, notable for its dual solubility in both water and fat. This unique property enables it to effectively neutralize a wide array of free radicals, including harmful species like superoxide and hydrogen peroxide, which can contribute to cellular damage and aging. Its broad-spectrum action makes ALA particularly beneficial in managing various health conditions, such as metabolic syndrome and diabetic neuropathy, where oxidative stress plays a significant role in disease progression. Furthermore, ALA has garnered attention for its promising effects in neurodegenerative diseases, as it not only helps reduce oxidative stress but also plays a crucial role in regenerating other antioxidants, such as vitamins C and E, enhancing the body's overall antioxidant defenses. By supporting mitochondrial function and promoting healthy glucose metabolism, ALA stands out as a potent ally in the prevention and management of chronic diseases associated with oxidative damage, thereby contributing to improved health outcomes and quality of life.

ALA exists in two mirror-image (isomeric) forms: the R-form (natural) and the S-form (synthetic). Supplements simply labeled as ALA usually contain a 50/50 mixture of R-ALA and S-ALA (called racemic). Research consistently shows that the R-form is more bioactive and effective at scavenging free radicals and protecting cells from oxidative stress.[186] It also has higher absorption and bioavailability, estimated at around 30% in humans.[187] Regular R-ALA is unstable and heat sensitive, undergoes extensive degradation during liver metabolism, and suffers from gastric instability, significantly decreasing its absorption, which is why it is not often commercially available. Stabilized forms, often created by complexing it with sodium, reduce these limiting factors and improve stability and absorption. Bioenhanced sodium R-lipoate ALA (Na-RALA) benefits from significantly increased bioavailability (2 to 3 times), with some claims suggesting up to 40-fold improvement.[188,189] Because of ALA's short half-life, smaller and more frequent dosing is recommended, with studies indicating improved absorption when taken on an empty stomach or in liquid form.[190,191]

Oxidative stress is closely associated with the pathogenesis of type 2 diabetes mellitus and its various complications. Several meta-analyses have examined the effects of ALA supplementation on cardiometabolic risk factors in individuals with this condition, with an emphasis on the role of ALA in individuals with type 2 diabetes and metabolic syndrome. It enhances insulin sensitivity, regulates blood glucose levels, and supports mitochondrial function. The conclusions of the meta-analyses are that ALA supplementation (typically 300–1200 mg/day) significantly reduces fasting blood glucose (the amount of sugar in the blood after not eating), HbA1c (a long-term marker of average blood sugar control over several months), C-reactive protein or CRP (a marker of inflammation in the body), and triglycerides (a type of blood fat linked to heart disease risk).[192,193] Interestingly, the most notable reduction in HbA1c occurred at 300 mg/day, while doses above 600 mg/day supported sustained weight loss. However, the clinical significance of these improvements remains modest, as many outcomes did not meet thresholds for meaningful clinical relevance.

While ALA significantly reduced fasting blood sugar and HbA1c in people with diabetes, it did not consistently improve insulin levels or HOMA-IR scores (Homeostatic Model Assessment of Insulin Resistance; a widely used method to assess insulin resistance), suggesting its glucose-lowering effects may operate independently of insulin sensitivity in some cases. ALA might directly help cells take in more sugar from the blood (by stimulating glucose transporter 4 and activating AMP-activated protein kinase and p38 mitogen-activated protein kinase), similar to how insulin works but through different pathways.[194] It also acts as a strong antioxidant, fighting off harmful stress in the body that can worsen diabetes, and might improve how our cells use energy. So, ALA seems to tackle high blood sugar through various routes, not just by fixing insulin resistance directly.

Oxidative stress is recognized as a significant contributor to the development of hypertension and other cardiovascular diseases. A meta-analysis investigated the effects of ALA supplementation on blood pressure in adults.[195] This study systematically reviewed and performed a dose-response meta-analysis of 11 randomized controlled trials involving 674 participants. The analysis, which also utilized the GRADE assessment to evaluate the certainty of evidence, revealed that ALA supplementation significantly reduced both systolic blood pressure (-5.46 mmHg) and diastolic blood pressure (-3.36 mmHg).

Subgroup analyses indicated that the reduction in both SBP and DBP was more pronounced in trials using ALA dosages of less than 800 mg/day and with a duration of ≤12 weeks. However, a non-linear dose-response analysis did not find a significant effect of ALA dosage or treatment duration on blood pressure. The certainty of evidence, according to the GRADE assessment, was moderate for SBP and high for DBP. These findings suggest that ALA supplementation could have a favorable impact on blood pressure levels in adults, with a clinically significant reduction observed in systolic blood pressure.

Inflammation, often intertwined with oxidative stress, is a key factor in the development and progression of many chronic diseases. Meta-analyses have examined the impact of ALA supplementation on various inflammatory markers in humans. A systematic review and meta-analysis aimed to determine the effects of ALA on the inflammatory markers TNF-α, CRP, and IL-6.[196] This analysis included nineteen studies from randomized controlled trials that investigated the effects of ALA on these inflammatory markers. The results indicated that ALA supplementation—300-1200 mg/day for 2 to 48 weeks—led to a statistically significant decrease in serum levels of CRP, IL-6, and TNF-α. These findings suggest a potential decreasing effect of ALA on key pro-inflammatory mediators, particularly when administered at higher doses. They also emphasized the need for more research with different intervention durations and considering potential sex-specific effects.

Another meta-analysis assessed the effect of ALA supplementation—300-1200 mg/day for 2 to 192 weeks—on both glycemic and inflammatory parameters, analyzing data from 41 randomized controlled trials.[197] Regarding inflammation, this study also found that ALA significantly reduced the levels of TNF-α, IL-6, and CRP. Subgroup analysis in this meta-analysis indicated that the reduction in CRP was more pronounced in trials with higher methodological quality and in participants without type 2 diabetes. The authors suggested that ALA's anti-inflammatory effects might be mediated through the inhibition of pro-inflammatory cytokines, possibly by interfering with the NF-κB signaling pathway. The consistent findings across these two independent meta-analyses provide strong evidence for the anti-inflammatory properties of ALA in humans. The observation that higher doses might be more effective warrants further research into optimal dosing strategies. How well ALA reduces inflammation (measured by CRP) can depend on the quality of the study and who is taking it, meaning it might work better for some people or in certain situations.

ALA possesses iron-chelating properties, which has led to investigations into its potential effects on iron metabolism. A meta-analysis evaluated the impact of ALA supplementation on iron homeostasis-related markers, including serum iron, total iron binding capacity (TIBC), hemoglobin, and ferritin.[198] This analysis included 10 randomized controlled trials with a total of 529 participants. The overall pooled analysis did not reveal any statistically significant effects of ALA supplementation on ferritin, serum iron, hemoglobin, or TIBC. However, subgroup analysis indicated that ALA supplementation significantly increased hemoglobin levels in people with hematological disorders—anemia, bleeding disorders, clotting disorders—and in studies with a duration longer than 8 weeks. These findings suggest that while ALA supplementation may not have a broad impact on iron metabolism in the general population, it could potentially benefit specific subgroups, such as individuals with hematological disorders or those receiving longer-term supplementation, by increasing hemoglobin levels.

One area of research focuses on the potential of ALA to protect the nervous system from oxidative damage. Sevoflurane, a commonly used inhaled anesthetic, has been shown to potentially induce oxidative stress in the brain. A human clinical trial investigated whether intravenous administration of 600 mg of ALA could mitigate this effect.[199] The study involved 155 individuals undergoing liver resection surgery with sevoflurane anesthesia, who were randomly assigned to receive either ALA or a placebo. The researchers measured oxidative stress markers (8-OHdG, sORP, and cORP) and brain injury biomarkers (S100β and UCH-L1) in perioperative blood samples.

The findings of this study indicated that ALA administration significantly reduced the levels of both oxidative stress and brain injury biomarkers induced by sevoflurane anesthesia. Furthermore, the patients who received ALA showed improved postoperative cognitive function compared to the placebo group, suggesting a neuroprotective effect. The study authors explained that sevoflurane can interfere with the energy-producing process in cells (mitochondrial electron transport chain), causing the production of excessive ROS and subsequent oxidative damage to cells. ALA, being a potent antioxidant, can effectively scavenge these ROS and promote the body's endogenous antioxidant defenses. Preclinical studies have also shown that ALA can cross the blood-brain barrier—meaning it can enter and affect the brain— and help protect nerve cells by turning on the body's internal antioxidant defense system— nuclear factor erythroid 2-related factor 2 (Nrf2) pathway—prompting cells to produce more of their own protective enzymes.[200,201,202] This single study provides initial evidence suggesting that ALA may play a significant role in reducing the neurological risks associated with a widely used anesthetic. The administration of a single intravenous dose of 600 mg of ALA prior to operating aligns with pharmacokinetic data suggesting that a high peak plasma concentration is crucial for its therapeutic efficacy in the brain. This highlights the importance of considering the route and timing of ALA administration to optimize its potential benefits, particularly in acute clinical settings.

ALA is well-known for its efficacy in treating neuropathic pain associated with diabetes. However, research also suggests its potential in managing other types of pain. A randomized controlled trial (RCT) evaluated the safety and effectiveness of ALA in reducing chronic pain in adults with normal blood sugar and suffering from various conditions, including neuropathic pain, arthralgia, and idiopathic myalgia.[203] The findings from this study indicated that oral ALA supplementation at dosages of 400 or 800 mg per day for two months significantly reduced the intensity of chronic pain, regardless of the underlying cause. This suggests that ALA may possess broader analgesic properties beyond diabetic neuropathy and could be beneficial for individuals experiencing chronic pain from various etiologies.

Chemotherapy, while still the standard treatment for many types of cancer, often leads to significant side effects, including peripheral neuropathy and increased oxidative stress. An RCT investigated the effects of ALA supplementation in women with breast cancer undergoing chemotherapy with doxorubicin and paclitaxel.[204] This study found that supplementation with 600 mg of ALA per day for six months during chemotherapy significantly improved peripheral sensory neuropathy and reduced levels of certain inflammatory and oxidative stress markers in these women. This suggests that ALA might have a supportive role in cancer therapy by helping to mitigate some of the adverse effects of chemotherapy, potentially improving quality of life among those undergoing invasive chemotherapy.

ALA has been shown to significantly reduce oxidative stress markers in people with diabetes, even those with poor glycemic control and albuminuria. Albuminuria is the presence of albumin (a type of protein) in the urine. It is a sign that the kidneys are not functioning properly, as healthy kidneys typically prevent significant amounts of protein from passing into the urine. Albuminuria is often an early indicator of kidney disease, particularly in conditions like diabetes, hypertension, and chronic kidney disease (CKD). People in the study took 600 mg daily for greater than three months. This suggests that ALA can improve the balance between oxidative stress and antioxidant defense in challenging clinical scenarios.[205] In a study on older adults with type 2 diabetes, ALA supplementation at 600 mg/day for six months showed a decrease in oxidative stress markers and inflammation.[206] However, the effects were similar to those of a placebo, indicating that higher doses might be necessary to achieve significant benefits.

A systematic review and meta-analysis evaluated the impact of ALA supplementation on oxidative stress parameters.[207] The analysis included 15 studies and found that ALA significantly decreased malondialdehyde (MDA) levels, indicating a reduction in lipid peroxidation. However, no significant effects were observed on other antioxidant enzymes and oxidative stress parameters, suggesting that ALA's primary benefit may be in controlling lipid peroxidation.

Alpha-lipoic acid, particularly in its R-form, stands out as a potent and versatile antioxidant with broad applications across metabolic, neurological, cardiovascular, and inflammatory conditions. Its unique dual solubility, ability to regenerate other antioxidants, and capacity to improve mitochondrial function make it a valuable therapeutic agent. While evidence supports its use in specific scenarios—such as diabetic neuropathy, blood pressure management, and adjunctive cancer care—its clinical efficacy often depends on the dose, route of administration, and the health status of the individual. Continued research is needed to refine optimal dosing strategies and identify populations that may benefit most from ALA supplementation.

Coenzyme Q10 (CoQ10)

Preferred Form(s): Water-soluble CoQ10 or ubiquinol | *Typical Adult Dosage*: Maintenance—50–200 mg/day; Intensive—300–3600 mg/day (500 mg commonly used), divided into three doses | *Companion Nutrients*: B vitamins, tyrosine, L-carnitine, phenylamine, S-adenosylmethionine, vitamin E, vitamin C | *When to Take*: Morning or midday with food

Coenzyme Q10 (CoQ10) is an essential compound that plays a critical role in mitochondrial bioenergetics, acting as a key component in the production of adenosine triphosphate (ATP), the primary energy currency of cells. It comes in two main forms: ubiquinone (the oxidized form) and ubiquinol (the reduced, active antioxidant form). Both are available as supplements, but the body must convert ubiquinone to ubiquinol to utilize it and the ability to do so decreases as we age. More recently, a water-soluble form has been created (Q10Vital) that may be four times more bioavailable.[208],[209] CoQ10 also exhibits potent antioxidant properties, effectively neutralizing lipid peroxyl radicals and other neutral radicals that can lead to cellular damage. Given its dual role in energy production and antioxidant protection, CoQ10 is particularly beneficial for maintaining heart health. Research indicates that CoQ10 levels naturally decline as we age, and notably, concentrations are often found to be significantly reduced in individuals suffering from heart disease. This depletion can impair mitochondrial function, further exacerbating issues related to cardiac health.

In addition to its cardiovascular benefits, CoQ10 supports overall cellular health and energy production across various tissues in the body. By enhancing mitochondrial function, it may improve energy availability in muscle cells, thereby supporting physical performance and recovery.[210] Furthermore, emerging evidence suggests that CoQ10 may have positive effects on neurodegenerative conditions, such as Alzheimer's and Parkinson's diseases, where oxidative stress and mitochondrial dysfunction play pivotal roles in disease progression.[211] Studies indicate that CoQ10 may help to stabilize mitochondrial membranes, improve neuronal energy metabolism, and reduce oxidative damage, potentially leading to better cognitive outcomes and quality of life for affected individuals. Moreover, the supplementation of CoQ10 has been linked to improvements in various health parameters, including blood pressure regulation and inflammatory markers, further emphasizing its role as a versatile agent in managing chronic conditions.[212, 213]

Several systematic reviews and meta-analyses have evaluated the impact of CoQ10 on oxidative stress biomarkers. CoQ10 supplementation has been associated with increased total antioxidant capacity (TAC) and reduced malondialdehyde (MDA) levels, indicating a reduction in oxidative stress.[214, 215, 216] However, the effects on other markers like SOD and CAT are less consistent, suggesting the need for further research to confirm these findings. Some of the studies show that CoQ10 is most effective in people with coronary artery disease or type 2 diabetes, which makes sense because diabetes dramatically increases oxidative stress and CoQ10 is highly concentrated in the heart.

A review examining the clinical implications of CoQ10 in the treatment of chronic disease and its effects on oxidative stress concluded that it shows promise in managing conditions driven by oxidative stress.[217] Conditions highlighted included cardiovascular diseases (like heart failure and coronary artery disease), kidney disease, chronic obstructive pulmonary disease (COPD), non-alcoholic fatty liver disease (NAFLD), and neurodegenerative diseases such as Alzheimer's and Parkinson's. In heart conditions, CoQ10 not only acts as an antioxidant but also improves energy production in heart cells and helps blood vessels function better. For kidney disease, CoQ10 supplementation has been shown to reduce oxidative stress and improve kidney function. In COPD, CoQ10 and related compounds have helped reduce inflammation and protect lung cells from damage caused by smoking. For NAFLD, CoQ10 may help reduce liver fat accumulation and inflammation. In neurodegenerative diseases, while research is ongoing, CoQ10 shows potential in protecting brain cells from damage. Essentially, CoQ10 is indispensable for cellular energy production, and therefore critical for optimizing cellular function, thereby enhancing the performance of all tissues and organs throughout the body.

Its ability to bolster the body's natural antioxidant defenses while simultaneously enhancing cellular energy production underscores CoQ10's significance not only in heart health but also in overall wellness. Beyond cardiovascular and kidney health, CoQ10 has been explored for its potential benefits in neurodegenerative diseases, migraine, cancer, and muscle disorders. Continued research into this remarkable coenzyme may unlock new avenues for intervention in a range of age-related diseases and conditions, ultimately promoting longevity and a higher quality of life.

Glutathione

Preferred Form(s): S-acetyl glutathione, liposomal glutathione, or sublingual glutathione | *Typical Adult Dosage*: Liposomal/S-acetyl glutathione—100–500 mg/day; Sublingual—150 mg three times daily | *Companion Nutrients*: N-acetyl cysteine, cysteine, selenium, vitamin B2, alpha-lipoic acid | *When to Take*: Morning or midday with or without food

Glutathione (GSH) is frequently hailed as the "master antioxidant" due to its exceptional capability to neutralize a diverse range of free radicals, including particularly aggressive species such as hydroxyl radicals and peroxides. This remarkable tripeptide—composed of three amino acids: cysteine, glutamine, and glycine—plays a pivotal role in maintaining cellular health and integrity. One of its primary functions is detoxification; glutathione actively binds to potentially harmful substances and helps convert them into less toxic forms that can be easily eliminated from the body, particularly through the liver. Its detoxifying properties are especially crucial in protecting against oxidative stress, which is associated with a plethora of chronic health conditions, such as neurodegenerative disorders like Alzheimer's and Parkinson's diseases, as well as liver diseases that can compromise overall metabolic function.

Oral standard glutathione supplementation has several limitations that limit its absorption, bioavailability, and effectiveness. Glutathione is broken down in the gastrointestinal tract by enzymes (such as γ-glutamyltransferase) before it can be absorbed into the bloodstream, reducing its bioavailability and effectiveness. Estimates suggest the bioavailability of oral glutathione is less than 1%.[218] Stomach acid and digestive enzymes degrade glutathione into its precursor amino acids (glutamate, cysteine, and glycine), limiting the direct absorption of intact glutathione. Even if some glutathione is absorbed in the intestines, cells have difficulty directly taking up glutathione from the bloodstream. Instead, they primarily rely on synthesizing glutathione from its precursors. Glutathione is prone to oxidation, and exposure to air or heat can reduce its potency, leading to a decrease in effectiveness over time. Glutathione has a short half-life in the body, meaning it is quickly metabolized and excreted, requiring frequent dosing to maintain adequate levels.

To overcome these limitations, two novel supplement forms have been developed: S-acetyl glutathione and liposomal glutathione. S-acetyl glutathione is a modified form of glutathione with an acetyl group attached, which enhances stability and the ability to pass through the GI tract. This form also has the ability to cross the blood-brain barrier, making it a good option for neurodegenerative conditions involving oxidative stress. Interestingly, a study found that supplementation with S-acetyl glutathione resulted in significantly higher plasma glutathione levels, but the standard form was slightly better absorbed by red blood cells (erythrocytes).[219] Erythrocytes have specific transport systems for glutathione. While S-acetyl glutathione is designed for systemic absorption and then intracellular conversion, it's possible that the standard glutathione might have a slightly more direct or efficient uptake mechanism into erythrocytes through their existing transporters. S-acetyl glutathione is considered to have better bioavailability than standard glutathione, allowing for increased cellular uptake. Interestingly, research notes that S-acetyl glutathione itself was not quantifiable in any plasma sample.[220] This suggests that its deacetylation (removal of the attached acetyl group) into active glutathione occurs very rapidly, likely during or immediately after absorption, making it a very effective prodrug. Liposomal encapsulation protects glutathione from breakdown in the digestive system as well. This leads to significantly improved absorption and delivery into cells. Studies suggest that liposomal delivery can substantially increase plasma glutathione levels compared to standard oral glutathione.[221, 222]

Another way to overcome these limitations is to use sublingual glutathione. Comparison of standard oral glutathione to N-acetyl cysteine (NAC; a glutathione precursor) and sublingual glutathione found that the sublingual form increased plasma levels to the greatest degree and was the only supplement that increased vitamin E levels as well.[223] This suggests that sublingual glutathione is superior to NAC and standard oral glutathione in both bioavailability and efficacy.

In addition to its detoxifying abilities, glutathione is integral to supporting immune function. It enhances the activity of immune cells, including lymphocytes and natural killer (NK) cells,[224, 225] which are essential for the body's defense against pathogens and the development of immune tolerance. Moreover, by reducing oxidative stress, glutathione contributes to a more balanced inflammatory response, thus preventing chronic inflammation often linked to various chronic diseases.

Beyond its role in detoxification and immunity, glutathione is essential for cellular repair mechanisms. It plays a vital role in the synthesis and repair of DNA, protein folding, and the recycling of other antioxidants, such as vitamins C and E, thereby amplifying the body's overall antioxidant defenses. As we age or face increased oxidative challenges—from poor diet, environmental toxins, or chronic stress—our glutathione levels can decline, potentially leading to diminished cellular function and heightened disease susceptibility.

A review highlighted the role of GSH in managing oxidative stress and inflammation in type 2 diabetes (T2D).[226] GSH supplementation, potentially combined with vitamin D3, may offer dual benefits by reducing oxidative stress and inflammation, thus improving insulin sensitivity and reducing complications associated with T2D. One randomized clinical trial supplemented people with diabetes with 500 mg oral standard glutathione daily for six months.[227] Blood levels of glutathione increased, and oxidative markers (8-OHdG) significantly decreased after glutathione supplementation. Additionally, fasting glucose and HbA1c improved in those over 55 years of age. The results indicate that oral glutathione may reduce complications associated with oxidative stress in people with diabetes, particularly those over age 55.

Chronic fatigue syndrome (CFS) is associated with oxidative stress, which contributes to its symptoms. In people with CFS, cellular mitochondria don't function properly, leading to increased free radical production. These free radicals can damage various parts of cells, including DNA and lipids, contributing to CFS symptoms. Oxidative stress can also trigger the production of inflammatory substances called cytokines, creating a chronic inflammatory response. Lastly, a particularly harmful molecule called peroxynitrite is often elevated in people with CFS. This molecule can further damage mitochondria, cause lipid peroxidation, and increase cytokine levels, creating a vicious cycle of oxidative stress and inflammation.

A study explored how various antioxidants—glutathione, N-acetylcysteine, alpha-lipoic acid, oligomeric proanthocyanidins, ginkgo biloba, and bilberry—can help reduce this oxidative stress and CFS symptoms.[228] It concluded that supplementing with glutathione, N-acetylcysteine, and alpha-lipoic acid can help the body produce more of its own antioxidants. Other antioxidant supplements—such as ginkgo biloba, bilberry, and oligomeric proanthocyanidins—can directly neutralize harmful molecules and protect cells from damage. Among these, oligomeric proanthocyanidins may be particularly effective because they significantly protect against DNA damage and lipid peroxidation caused by free radicals. By increasing endogenous antioxidants and taking antioxidant supplements, oxidative stress can be reduced, leading to a reduction in CFS symptoms.

One study concluded that oral supplementation with glutathione in healthy adults did not significantly change oxidative stress markers.[229] 40 healthy volunteers took 500 mg of oral standard glutathione, twice daily for four weeks. Reductions in oxidative stress markers, creatinine, and serum glutathione were not changed after supplementation. This may be due to poor bioavailability of standard glutathione as well as limited oxidative stress in healthy individuals.

Inhaled glutathione has also been investigated as a treatment for neurodegenerative conditions. Preclinical research on nebulized glutathione suggests it as a promising method for delivering antioxidants directly to the central nervous system.[230] This approach may enhance bioavailability and therapeutic efficacy, potentially offering neuroprotection and mitigating oxidative stress in neurodegenerative diseases like Parkinson's and Alzheimer's.

One last form to mention is topical stabilized glutathione, which has garnered significant interest in dermatology due to its antioxidant, anti-inflammatory, and melanin-regulating properties. This topical glutathione is stabilized through liposomal encapsulation, acetylation, including precursors (cysteine, glutamic acid, glycine), or by combining with other antioxidants. While research focuses on its use for skin brightening, evening the skin tone, maintaining youthful skin appearance, and inflammatory skin conditions (eczema, acne, psoriasis), topical stabilized glutathione may also be useful for shingles outbreaks and neuropathic pain. While clinical research is lacking currently for these uses, anecdotal and clinical experience suggests it is helpful.

Research continues to explore the benefits of glutathione supplementation and strategies for boosting its levels within the body, given its foundational importance in health and longevity. By ensuring optimal levels of this powerful antioxidant, individuals may be better equipped to combat oxidative stress and its associated health risks, thereby promoting improved overall well-being and resilience against illness.

Resveratrol and Pterostilbene

Preferred Form(s): Pterostilbene; Liposomal or nanoemulsified trans-Resveratrol | *Typical Adult Dosage*: Pterostilbene—50–150 mg/day, 250 mg/day for intensive use; Liposomal—100–300 mg/day; Nanoemulsion—50–250 mg/day; Standard—250–500 mg, up to four times daily | *Companion Nutrients*: Vitamin C, vitamin D, curcumin, quercetin, grapeseed extract, nicotinamide riboside, nicotinamide mononucleotide (NMN) | *When to Take*: Resveratrol—Morning to early evening with food; Pterostilbene—Morning to early evening with or without food

Resveratrol is a powerful polyphenolic compound predominantly found in the skins of red grapes, as well as in other sources such as berries, peanuts, and certain herbs. Most supplements are derived from Japanese knotweed (*Reynoutria japonica*, syn: *Polygonum cuspidatum*). Its bioactive properties stem from its ability to act as an antioxidant, effectively neutralizing harmful peroxyl radicals and reactive nitrogen species (RNS). These damaging agents are implicated in a range of physiological conditions, including oxidative stress, chronic inflammation, and the aging process. Numerous studies have demonstrated that resveratrol can impart cardiovascular benefits by enhancing endothelial function, which involves the health and operation of the inner lining of blood vessels.[231] Improved endothelial function is associated with better vascular relaxation and reduced inflammatory markers, ultimately contributing to a lower risk of heart disease. Additionally, emerging research suggests that resveratrol may play a role in metabolic syndrome—metabolic dysfunction characterized by a cluster of conditions that increase the risk of heart disease and diabetes—by improving insulin sensitivity and promoting healthy lipid metabolism.[232]

To further understand the benefits of resveratrol, it's vital to delineate the distinction between trans-resveratrol and cis-resveratrol. Resveratrol exists in two main geometric forms or isomers: trans-resveratrol and cis-resveratrol. Trans-resveratrol is the predominant and biologically active form, recognized for its robust anti-inflammatory and antioxidant effects. This form has been extensively studied for its potential health benefits, including its role in activating sirtuins,[233] proteins associated with longevity and metabolic regulation. Conversely, cis-resveratrol, which is less prevalent in nature, has significantly lower biological activity compared to its counterpart. Consequently, when considering resveratrol supplementation or consumption of resveratrol-rich foods, it is essential to prioritize sources that provide trans-resveratrol, as its therapeutic potential has garnered the most scientific support.

The poor bioavailability of resveratrol in humans due to its rapid and extensive metabolism in the liver and intestine makes it difficult to reach efficacious blood levels. Research indicates that its oral bioavailability is less than 1%.[234] About 75% of resveratrol is excreted in the feces and urine alone, with nearly all of the remaining resveratrol forming various metabolites as resveratrol glucuronides and resveratrol sulfates.[235] Researchers have attempted to overcome this limitation by combining it with other ingredients (e.g., piperine, quercetin), derivatizing it (chemically modify), or encapsulating it.[236,237] Derivatized forms do show greater stability and bioavailability in preclinical models. Similarly, encapsulation in nanoparticles increases its solubility in GI fluids, allowing higher absorption. Lastly, liposomal encapsulation improves solubility, stability, and bioavailability, enhancing its therapeutic activity.

Resveratrol helps the body reduce inflammation and protect against cell damage by activating a natural defense system called the Nrf2 signaling pathway.[238] Think of Nrf2 as a "master switch" that turns on genes responsible for fighting inflammation and oxidative stress (the damage caused by harmful molecules called free radicals). When resveratrol activates this switch, it helps the body produce more antioxidants and protective enzymes, reducing overall damage and promoting better health.

In type 2 diabetes, resveratrol helps reduce harmful cell damage and boost the body's natural defense system.[239] It does this by activating Nrf2, which leads to an increase in antioxidants like SOD. This helps neutralize harmful molecules and reduce stress on the body's cells. A meta-analysis of randomized controlled trials involving people with type 2 diabetes demonstrated that resveratrol supplementation significantly reduced levels of C-reactive protein, lipid peroxides, and oxidative stress scores.[240] It also increased levels of GPX and CAT, although the evidence level was low, indicating a need for further research.

Resveratrol has also been clinically investigated for ulcerative colitis (UC). UC is a condition where the immune system attacks the colon, causing inflammation and damage. A big part of this damage comes from oxidative stress, which happens when free radicals build up and attack healthy colon cells. In individuals with UC, resveratrol supplementation improved oxidative stress markers, reduced disease activity, and enhanced quality of life by increasing total antioxidant capacity and reducing malondialdehyde levels.[241] The participants consumed 500 mg of resveratrol for six weeks.

Overall, as research continues to explore the full extent of resveratrol's health benefits, its role in promoting cardiovascular health, managing metabolic disorders, and sustaining neurological function amplifies its importance within functional foods and nutraceutical development. If its bioavailability can be enhanced, it may prove to have a significant impact on human health.

Found naturally in blueberries, cranberries, grapes, almonds, and the bark of the Indian almond tree (*Terminalia catappa*) and heartwood of the Indian kino (*Pterocarpus marsupium*), pterostilbene is chemically related to resveratrol with one key difference that makes it more lipophilic and bioavailable. Pterostilbene has two methoxy groups (compared to resveratrol's hydroxyl groups), which makes it more lipid-soluble and helps it be absorbed better and stay in the body longer. This small change makes pterostilbene more bioavailable and effective at lower doses than resveratrol. Like resveratrol, pterostilbene has antioxidant and anti-inflammatory properties, and it activates the Nrf2 pathway, which helps protect against oxidative stress and inflammation.

A review explored the antioxidant properties of pterostilbene and its potential to prevent and treat various diseases.[242] Pterostilbene has shown promising effects in laboratory and animal studies, demonstrating benefits against cancer, neurological disorders, inflammation, vascular disease, and diabetes. The studies suggest that pterostilbene's antioxidant activity helps protect cells from damage and may contribute to its therapeutic potential.

Pterostilbene is better than resveratrol at protecting the cells in your intestines from damage caused by oxidative stress.[243] It helps the mitochondria work better, and even grow more when needed, by activating a special pathway in the body called SIRT1.[244] This pathway helps repair cells, boost energy production, and protect the intestines from damage. So, pterostilbene helps keep the mitochondria in intestinal cells healthy and functioning, which is important for preventing injury from oxidative stress.[245]

Clinical research on pterostilbene is currently limited, but some promising research is emerging. First of all, it has demonstrated safety and tolerability up to 250 mg/day.[246] Some studies using the high end of dosing have observed modest increases in both total and LDL cholesterol, but these can be mitigated by combining with grape extract. A randomized controlled trial evaluated pterostilbene in people with high cholesterol, reporting that 250 mg/day significantly reduced both systolic and diastolic blood pressure.[247] Furthermore, people not taking a cholesterol-lowering drug experienced minor weight loss while taking pterostilbene.[248] This is a promising finding for cardiovascular health. A six-month randomized, double-blind, placebo-controlled trial investigated a combination of nicotinamide riboside and pterostilbene in adults with non-alcoholic fatty liver disease (NAFLD). While no significant change was seen in fatty liver, reduced liver enzymes and liver inflammatory markers were observed, suggesting the combination improved liver function and reduced liver inflammation.[249] Human clinical research on pterostilbene, though still in its early stages, has established its general safety and tolerability. It has also shown some promising results, particularly in reducing blood pressure and potentially lowering markers of liver inflammation. The superior bioavailability of pterostilbene compared to resveratrol, which has been well-established in preclinical studies, provides a strong rationale for continued human research to fully explore its therapeutic potential, especially in areas like oxidative stress, inflammation, and gut health.

Pterostilbene emerges as a potent and promising compound, often regarded as a more bioavailable and effective alternative to resveratrol. Its unique chemical structure, featuring two methoxy groups, enhances its lipid solubility, which in turn facilitates better absorption and longer retention within the body. This increased bioavailability can potentially yield significant health benefits at lower doses, making it a compelling choice for those seeking to harness the advantages of natural antioxidants. With research highlighting its protective effects against a myriad of conditions—including cancer, neurological disorders, vascular diseases, and metabolic syndrome—pterostilbene's reputation in the health and wellness community continues to grow. Moreover, its ability to support mitochondrial health, particularly in intestinal cells, underscores its crucial role in combating oxidative stress and enhancing energy production. As scientific exploration uncovers further insights into its mechanisms of action and therapeutic potential, pterostilbene holds promise not only as a supplement but also as a valuable asset in preventative health strategies aimed at improving overall cellular function and resilience.

Curcumin

Preferred Form(s): Curcumin-galactomannan complex (CGC); curcumin-essential oil complex (CEOC); Lipid-encapsulated curcumin (LEC) | *Typical Adult Dosage*: CGC—250–500 mg once or twice daily; CEOC—300–500 mg, once or twice daily; LEC—500 mg up to twice daily | *Companion Nutrients*: Omega-3 fatty acids, quercetin, *Boswellia serrata* extract, B vitamins, vitamin C, vitamin E, resveratrol | *When to Take*: With meals throughout the day

The active polyphenol compound in turmeric, curcumin, is renowned for its substantial antioxidant activity, primarily achieved through its ability to scavenge free radicals such as ROS and RNS.[250] By neutralizing these free radicals, curcumin helps protect cellular integrity and maintain overall health. This protective effect is especially important in managing oxidative stress.

In addition to its antioxidant properties, curcumin exhibits potent anti-inflammatory effects, making it a valuable ally in the management of a variety of inflammatory conditions. By modulating key inflammatory pathways, such as the NF-kB signaling pathway and the production of pro-inflammatory cytokines, curcumin can help alleviate symptoms and slow the progression of diseases such as arthritis, where chronic inflammation leads to joint pain and damage.[251] Furthermore, curcumin's role in cardiovascular health cannot be understated;[252] it has been shown to improve endothelial function, preserve mitochondrial function in the heart, reduce levels of LDL cholesterol, and enhance overall vascular health, thereby lowering the risk of heart disease.

Curcumin's therapeutic potential extends to neurodegenerative disorders, including Alzheimer's disease.[253] Research suggests that curcumin can cross the blood-brain barrier, where it exerts protective effects on neuronal cells by reducing oxidative stress and inflammation in the brain.[254] It's also involved in the regulation of insulin pathways and glucose metabolism in the brain.[255] These actions are crucial, as dysregulated brain insulin sensitivity and glucose utilization alongside chronic inflammation are hallmarks of Alzheimer's pathology. By intervening in these processes, curcumin may help preserve cognitive function and delay neurodegeneration. Moreover, its ability to promote neurotrophic factors,[256] which support neuron health and growth, positions curcumin as a promising candidate for neuroprotective strategies.

The bioavailability of curcumin is generally considered low due to its poor absorption in the GI tract, rapid metabolism in the liver and intestines, and swift elimination.[257] To overcome this limitation multiple innovative formulations have been developed, but the three that appear to be most effective are curcumin-galactomannan complex (CGC; brand name CurQfen), curcumin-essential oil complex (CEOC), or a lipid encapsulated curcumin. According to studies commissioned by the manufacturer of CurQfen, combining curcumin with galactomannan fiber from fenugreek improves serum bioavailability up to 45.5-fold and enhances tissue distribution to the brain (an astonishing 245-fold increase compared to standard curcumin), heart, liver, kidney, and spleen.[258,259] CEOC improves the solubility of curcumin, its absorption through the intestinal wall, and limits its metabolic breakdown. Research shows that the bioavailability of this form is 6.93-fold higher than standard curcumin and 6.3-fold higher than curcumin with piperine.[260] Some research even suggests that complexing curcumin with turmeric essential oil can increase its bioavailability 7–10-fold compared to standard curcumin extract.[261,262,263,264] Additionally, the oil contains other bioactive molecules, especially *ar*-Turmerone, which provides additional benefits such as reduced neuroinflammation and the production of neural stem cells.[265,266] A third option to increase bioavailability would be curcumin in an advanced lipid matrix (Longvida). By encasing the curcumin in a lipid layer, it not only is protected from breakdown in the digestive system, but it also facilitates absorption into the lymphatic system, bypassing the liver's first-pass metabolism. This form is specifically designed to enhance delivery to the brain and increases bioavailability 65-fold.[267] Improved bioavailability with these forms of curcumin leads to increased therapeutic efficacy at lower doses.

A meta-analysis of randomized clinical trials assessing the antioxidant potential of curcumin concluded that it significantly elevates total antioxidant capacity and decreases malondialdehyde (MDA) concentration.[268] The average amount of curcumin consumed was 645 mg daily for an average of 67 days to achieve these results. The researchers suggests that curcumin is effective against oxidative stress because of its ability to remove ROS and RNS, act as a heavy metal chelator —meaning it can bind to and safely capture harmful metals so they can't trigger further damage—and regulate numerous enzymes.

Another meta-analysis assessed the effects of curcumin on biomarkers of inflammation and oxidative stress. Fifteen randomized controlled trials were included in the final analysis.[269] The researchers reported that curcumin significantly reduces both inflammation and oxidative stress as evidenced by reductions in IL-6, hs-CRP, and MDA. Based on these results, curcumin could be beneficial for cardiovascular diseases (atherosclerosis, heart attack, heart failure, stroke, peripheral artery disease), metabolic syndrome, type 2 diabetes, obesity, nonalcoholic fatty liver disease, arthritis, lupus, inflammatory bowel disease, systemic vasculitis, cancer, chronic kidney disease, allergic-related conditions (asthma, rhinitis, eczema), infections, and sepsis.

A third meta-analysis that including many studies (66 experiments in total) investigated if turmeric and its active ingredient, curcumin, help fight inflammation and oxidative stress in the body.[270] The results showed that taking turmeric or curcumin supplements significantly lowered several signs of inflammation (CRP, TNF-α, IL-6). They also improved the body's antioxidant defenses by increasing SOD and decreasing MDA, helping to protect cells. The study suggests that using turmeric or curcumin supplements might be a helpful way to reduce inflammation and boost antioxidant levels.

A fourth meta-analysis looked at whether curcumin can increase endogenous antioxidant production and reduce lipid peroxides.[271] After evaluating the results of seven clinical trials, the researchers determined that supplementing curcumin for six weeks or more significantly increases SOD, CAT, and GSH levels, and reduces lipid peroxides.

Overall, curcumin's multifaceted mechanisms—including its potent antioxidant and anti-inflammatory properties—underscore its significance as a promising natural remedy for a variety of health conditions. The research highlights its potential in combating oxidative stress, reducing inflammation, and offering neuroprotective benefits, which positions curcumin as a key player in the search for effective therapies against chronic diseases and age-related disorders. As we continue to uncover the complexities of its actions within the body, curcumin's therapeutic applications are becoming increasingly relevant, warranting further investigation to unlock its full potential in preventive and integrative medicine. With ongoing studies shedding light on innovative formulations that enhance its bioavailability, curcumin stands at the forefront of natural health interventions, offering hope for improved health outcomes in diverse populations.

Green Tea Extract (EGCG)

Preferred Form(s): EGCG phytosome or liposomal EGCG standardized to catechin/EGCG content | *Typical Adult Dosage*: Phytosome—150–300 mg, once or twice daily; Liposomal—100–500 mg/day | *Companion Nutrients*: Vitamin C, fish oil, quercetin, curcumin, resveratrol | *When to Take*: Morning or early afternoon with food

Epigallocatechin gallate (EGCG), a prominent polyphenolic compound found in green tea, has garnered significant attention for its robust antioxidant properties, which play a crucial role in combating oxidative stress and its associated diseases. EGCG is known for its ability to scavenge free radicals, effectively neutralizing these harmful species and preventing oxidative damage to cells and tissues. By donating electrons, EGCG mitigates the activity of hydroxyl radicals and superoxide anions, helping maintain cellular integrity and function. Additionally, EGCG has been shown to enhance the activity of various endogenous antioxidant enzymes, such as SOD and GPX,[272] and activates the Nrf2/ARE pathways which upregulates cell protective genes and antioxidant proteins,[273,274] further bolstering the body's defense against oxidative stress.

One of the most notable benefits of EGCG is its positive effect on cardiovascular health.[275] Oxidative stress is a key contributor to the development of atherosclerosis and other cardiovascular diseases. EGCG promotes endothelial function, which is crucial for maintaining vascular health.[276] It enhances nitric oxide availability, leading to improved vasodilation and reduced blood pressure.[277] Furthermore, its anti-inflammatory properties help to decrease the expression of adhesion molecules and inflammatory cytokines,[278] which are often elevated in conditions of oxidative stress.

EGCG also plays a role in weight management and metabolic health. Studies have suggested that EGCG can mildly enhance fat oxidation and improve metabolic rate, which may be particularly beneficial in combating obesity—a significant risk factor for metabolic syndrome.[279] However, the weight loss is typically small and may not reach clinical significance. By reducing inflammation and oxidative stress associated with excess adiposity, EGCG may help improve insulin sensitivity and regulate metabolic processes.

The role of EGCG in cancer prevention has been the focus of numerous studies, revealing its potential to inhibit the proliferation of cancer cells and induce apoptosis (programmed cell death).[280] By reducing oxidative stress, EGCG may mitigate DNA damage that could otherwise lead to carcinogenesis. This antioxidant's ability to regulate various signaling pathways involved in cell growth and survival contributes to its anticancer properties, making it a candidate for dietary approaches in cancer prevention strategies.

In addition to cardiovascular diseases and cancer, EGCG's role in addressing other oxidative stress-related conditions cannot be overlooked. As we've seen, chronic inflammatory diseases, neurodegenerative disorders like Alzheimer's and Parkinson's, and even diabetes have connections to oxidative stress. Research indicates that EGCG may possess neuroprotective qualities by reducing oxidative damage in neuronal cells, potentially offering a protective effect against cognitive decline.[281]

Moreover, EGCG's influence on gut microbiota composition can lead to enhanced nutrient absorption and greater production of beneficial short-chain fatty acids, contributing to lower systemic inflammation and oxidative stress.[282] This interplay underscores the importance of dietary antioxidants in maintaining overall health and mitigating chronic diseases.

While much of the research on EGCG remains theoretical and based on preclinical models and mechanistic understanding, it still demonstrates significant promise across a variety of conditions.

EGCG has been identified as a beneficial compound in managing nonalcoholic fatty liver disease (NAFLD), a condition heavily influenced by oxidative stress.[283, 284, 285] Preclinical trials and meta-analyses have demonstrated that EGCG can alleviate oxidative stress and related metabolic dysfunctions, inflammation, and fibrosis in NAFLD. The compound acts through various signaling pathways, including Nrf2, AMPK, and NF-κB, to maintain redox homeostasis and protect against disease progression. These findings underscore EGCG's potential as a promising therapeutic agent for NAFLD, warranting further clinical investigation given the condition's rapidly increasing global prevalence.

In respiratory diseases, EGCG acts as a strong antioxidant, scavenging reactive oxygen species and inhibiting pro-oxidant enzymes.[286, 287] It also activates antioxidant systems, which can be beneficial in conditions like acute respiratory distress syndrome, asthma, and chronic obstructive pulmonary disease. EGCG's anti-inflammatory and anti-fibrotic effects further contribute to its potential therapeutic role in these diseases. Therefore, EGCG's multifaceted actions as an antioxidant, anti-inflammatory, and anti-fibrotic agent underscore its significant promise as a therapeutic intervention for a range of respiratory diseases.

Oxidative stress plays a significant and detrimental role in both male and female reproductive health, contributing to infertility and adverse pregnancy outcomes. Specifically, oxidative stress damages sperm cells, reduces semen quality, damages oocytes, contributes to oocyte mitochondrial dysfunction, impedes embryonic development, negatively impacts endometrial tissue, and disrupts hormonal regulation. EGCG has shown promise in improving reproductive health by reducing oxidative stress, which can damage DNA, lipids, and proteins in the reproductive system.[288] This reduction in oxidative stress can enhance sperm motility and oocyte maturation, potentially improving fertility outcomes and overall reproductive health. The antioxidant properties of EGCG make it a valuable compound in managing oxidative stress-associated infertility.

A randomized clinical study looked at whether taking green tea extract (GTE) tablets could help adults with transfusion-dependent β-thalassemia, a condition affecting red blood cells.[289] 27 participants were divided into three groups: one group took a placebo (a pill with no active ingredient), another took a GTE tablet equal to 50 mg of the main active component EGCG, and the third group took a GTE tablet equal to 100 mg EGCG. They took one tablet daily for 60 days. The study found that people taking the GTE tablets, especially the 100 mg dose, showed improvements in their hemoglobin levels compared to those taking the placebo, particularly after 60 days. The GTE tablets also seemed to lower levels of erythroferrone (a substance involved in red blood cell production and iron control) and bilirubin (a substance produced when red blood cells break down), suggesting a potential positive effect on red blood cell health and turnover. Importantly, there was also a trend toward lower iron levels in red blood cell membranes among those taking GTE, which is considered beneficial in this population because excess iron can promote oxidative damage and worsen disease complications. Overall, the results suggest that these GTE tablets might help improve some blood-related issues in people with transfusion-dependent β-thalassemia.

Inconsistent results between *in vitro* (tested in cells in a lab) and *in vivo* (tested in living organisms) studies may be the result of low bioavailability of catechins like EGCG. Several forms of EGCG or green tea extract have been created to enhance its bioavailability and efficacy. One of the most well-researched is EGCG complexed with phospholipids (phytosome).[290,291] This combines phospholipids like sunflower lecithin with the EGCG to create a lipid-friendly structure, which protects the EGCG from degradation and facilitates greater bioavailability in the GI tract. Similar in principle to phytosomes, liposomal technology encapsulates EGCG within tiny lipid vesicles (liposomes). This protects the EGCG and can enhance its uptake into cells.[292] Other ways to enhance the bioavailability of EGCG is to co-administer it with certain substances, like piperine, vitamin C, or fish oil.

While often beneficial, EGCG, particularly in concentrated extract form, has been linked to liver injury in some individuals. This adverse effect is often idiosyncratic, meaning it's unpredictable and not directly dose-dependent in all cases, suggesting that individual susceptibility plays a crucial role. Research indicates that genetic predispositions can significantly influence this risk. For instance, specific variations in genes like UGT1A4 and COMT, which code for enzymes involved in EGCG breakdown, may lead to less efficient metabolism and a buildup of EGCG or its metabolites in the liver, increasing the potential for damage.[293,294] Furthermore, individuals with the HLA-B*35:01 allele have been found to have a significantly higher risk of green tea extract-induced liver injury, suggesting an immune-mediated mechanism in susceptible individuals.[295] It's also been observed that consuming EGCG supplements while on a diet or fasting might increase the risk of liver injury by altering lipid metabolism pathways.[296] Therefore, while green tea itself is generally safe, high-dose EGCG supplements can pose a risk, especially for those with certain genetic risks.

In summary, EGCG from green tea serves as a powerful ally in the fight against oxidative stress and its myriad of associated diseases. Its multifaceted mechanisms—from direct scavenging of free radicals to improving endothelial function, reducing inflammation, and promoting metabolic health—highlight the importance of integrating this potent antioxidant into dietary habits. As research continues to illuminate the connections between oxidative stress and chronic disease, EGCG emerges as an invaluable component in strategies aimed at enhancing health and preventing disease, paving the way for more focused dietary recommendations and therapeutic interventions.

Selenium

Preferred Form(s): Selenium-enriched yeast (SEY) or selenomethionine | *Typical Adult Dosage*: SEY—200-285 mcg/day; selenomethionine—200 mcg/day | *Companion Nutrients*: Vitamin E, vitamin C, B vitamins, zinc, magnesium, fish oil, cysteine, glycine | *When to Take*: Anytime during the day with food

Selenium is a vital trace mineral that significantly contributes to human health primarily through its incorporation into selenoproteins, many of which serve as antioxidants. Selenoproteins are a group of proteins that contain the essential trace element selenium in the form of the amino acid selenocysteine, which is often referred to as the 21st amino acid. When our body makes proteins, it can add selenocysteine at specific spots, and this addition is essential for the proteins to work correctly. These proteins are vital for various physiological processes, including antioxidant defense, redox signaling, thyroid hormone metabolism, and immune function. In simple terms, selenoproteins are essential proteins that rely on selenium to perform a variety of crucial tasks in our body, allowing us to stay healthy and function well.

Selenium exists in several forms, including selenomethionine, sodium selenite, and sodium selenate, each with varying bioavailability and biological activity. Selenomethionine, the most common dietary form found in many plant-based foods, acts as a significant storage form of selenium in tissues, while inorganic forms like sodium selenite and sodium selenate are often used in supplements. The bioavailability of selenium can be influenced by dietary factors, individual differences such as age and gut health, and the specific forms of selenium consumed, with organic forms generally having better absorption rates compared to inorganic ones.

Selenomethionine is considered highly bioavailable (> 90%) because it is the organic form of selenium and readily incorporated into proteins in place of methionine. Research also suggests that it has a longer retention in the body and tends to increase selenium plasma levels more efficiently compared to inorganic forms.[297] Selenium-enriched yeast contains a mixture of selenium compounds, predominantly selenomethionine, along with other selenium-containing amino acids. Its bioavailability is generally comparable to selenomethionine, averaging between 74% and 89% bioavailability. Sodium selenite is absorbed more rapidly—greater than 50% absorption—but also excreted more quickly, meaning it has high bioavailability but poor retention. Its bioretention is higher than that of the selenate form. Given it is also an inorganic form, sodium selenate is considered comparable to sodium selenite. Research suggests it is 100% absorbed, but a significant amount of it is excreted in the urine. Overall, organic forms like selenomethionine and selenium-enriched yeast tend to have higher bioavailability and better tissue retention compared to inorganic forms such as selenite and selenate. This makes them more effective for long-term selenium supplementation.

Selenium's antioxidant defense primarily involves its incorporation into the enzyme GPX, which catalyzes the reduction of hydrogen peroxide to water, thereby preventing oxidative damage to cells. Additionally, selenium-containing enzymes can reduce lipid peroxides, protecting polyunsaturated fatty acids in cell membranes from oxidative stress. Adequate selenium levels are associated with a reduced risk of certain cancers—such as prostate, colorectal, and lung cancers—likely due to its role in enhancing antioxidant defenses and regulating cellular processes like apoptosis and DNA repair.[298] Furthermore, selenium may contribute to cardiovascular health by reducing oxidative stress and inflammation, helping to protect low-density lipoproteins (LDLs) from oxidation, which is a critical step in the development of atherosclerosis.[299]

Selenium is also crucial for optimal immune function, enhancing the proliferation of immune cells, particularly T lymphocytes, and modulating inflammation, which may help reduce the risks of immune-mediated diseases.[300] Emerging research suggests that selenium may have neuroprotective effects, potentially lowering the risk of neurodegenerative diseases such as Alzheimer's and Parkinson's by protecting neurons from oxidative damage.[301] It also plays a crucial role in supporting thyroid health by participating in the metabolism of thyroid hormones.[302] It is a key component of the enzyme deiodinase, which converts the inactive thyroid hormone thyroxine (T4) into its active form, triiodothyronine (T3). This conversion is essential for regulating metabolism and energy levels in the body. Additionally, adequate selenium levels can help reduce excess production of thyroid-stimulating hormone (TSH), which is released by the pituitary gland to stimulate the thyroid. High levels of TSH can indicate thyroid dysfunction, so maintaining optimal selenium levels may support overall thyroid function and contribute to hormonal balance. By promoting efficient thyroid hormone production and regulation, selenium helps maintain a healthy metabolism and prevents conditions such as hypothyroidism and Hashimoto's.

Selenium has been identified as a promising agent in managing oxidative stress-related ocular conditions such as cataracts, age-related macular degeneration, and diabetic retinopathy.[303] Its role as a component of selenoproteins and antioxidant enzymes helps counteract reactive oxygen species, potentially slowing the progression of these chronic disorders. Recent advancements in delivery and nanotechnology have enhanced the efficacy of selenium-based treatments for ocular health.

Laboratory research illustrates selenium's potential in protecting cardiomyocytes (muscle cells that make up the heart) from oxidative stress-induced damage, particularly by activating the PI3K/AKT signaling pathway.[304] This activation helps alleviate cell cycle arrest and promotes cell cycle progression, which is crucial in preventing heart failure. Essentially, activation encourages cells to continue their normal cycle of growth and division, preventing them from getting stuck and stopping their activity. When heart cells can divide and grow properly, the heart functions better and is less likely to develop heart failure. The study highlights selenium's role in enhancing the antioxidant capacity of selenoproteins like glutathione peroxidase and thioredoxin reductase, which are vital for cardioprotection.

A clinical trial demonstrated that selenium supplementation significantly reduced oxidative stress markers and improved clinical symptoms in people who suffer from migraines.[305] Participants in the trial took either 200 mcg of selenomethionine daily or a placebo (200 mcg starch) for 12 weeks. The study found that selenium reduced nitric oxide levels and increased total antioxidant capacity. It also decreased headache frequency and severity, suggesting selenium as a complementary therapy for migraine management.

Selenium supplementation, combined with diet therapy, has been effective in managing oxidative stress in individuals with metabolic syndrome.[306] This approach has shown promise in reducing oxidative stress markers and improving clinical parameters, suggesting selenium's potential in managing metabolic health.

Interestingly, a comparison of the protective effects of two forms of selenium— selenium-enriched yeast (SEY) and selenomethionine—against prostate cancer, concluded that SEY reduces biomarkers of oxidative stress, but selenomethionine does not.[307] SEY is produced by cultivating yeast in selenium-rich conditions, and contains a mix of organic selenium compounds, primarily selenomethionine, along with other forms of selenium, such as selenocysteine and lower proportions of other selenoamino acids. The presence of various selenium compounds may provide a broader spectrum of activity and enhance the overall antioxidant potential. Participants in the SEY group took either a low dose (200 mcg SEY) or high dose (285 mcg SEY) daily, while those in the selenomethionine group took 200 mcg daily. While plasma levels of selenium increased 93% in the selenomethionine group compared to 54% and 86% in the low- and high-dose SEY groups respectively, the greatest reductions in oxidative stress markers were observed in the high-dose SEY group. These findings highlight that SEY may have greater antioxidant activity, making it more effective in the reduction of oxidative stress despite lower plasma levels.

A comprehensive review of selenium supplementation has shown its effectiveness in reducing oxidative stress biomarkers, such as malondialdehyde, while increasing glutathione and total antioxidant capacity.[308] These findings support selenium's role in managing oxidative stress-related conditions.

Overall, selenium is an essential mineral with potent antioxidant capabilities mediated largely through its role in selenoproteins. Its various forms exhibit differing levels of bioavailability, impacting how effectively the body can utilize it. Given its role in mitigating oxidative stress, selenium supplementation could be a valuable strategy for reducing the risk of several health conditions linked to oxidative damage and inflammation. However, individual dietary sources and selenium amounts should be monitored to avoid potential toxicity, as excessive selenium intake can lead to selenosis, a toxic condition resulting from selenium overload.

Zinc

Preferred Form(s): Zinc bisglycinate | *Typical Adult Dosage*: Zinc bisglycinate maintenance—15–60 mg/day; Zinc bisglycinate intensive—50 mg, three times per day, when recommended by a healthcare professional and for up to 12 weeks | *Companion Nutrients*: Copper, vitamin C, vitamin A, B vitamins, quercetin | *When to Take*: Anytime during the day with food

Zinc is an essential trace mineral that plays a vital role in our body, particularly when it comes to protecting cells from oxidative stress. As a cofactor for various antioxidant enzymes, zinc helps these enzymes function effectively, assisting in the neutralization of free radicals, specifically superoxide and hydroxyl radicals. These free radicals can cause damage to cells, leading to increased oxidative stress, which has been linked to a variety of health issues, including age-related diseases such as Alzheimer's and other neurodegenerative conditions, weakened immune function, and chronic inflammatory states.

When there is not enough zinc in the body—called zinc insufficiency or deficiency—the antioxidant defense weakens, allowing oxidative stress to increase. This is particularly concerning as oxidative stress can lead to the malfunction of cells and tissues, contributing to the progression of diseases and overall aging. On the other hand, maintaining adequate levels of zinc supports not only immune health but also promotes skin integrity, playing a crucial role in wound healing and managing skin disorders like acne and dermatitis.

Quercetin is particularly valuable for improving zinc absorption because it acts as a zinc ionophore. As such, it may help facilitate the transport of zinc across cell membranes, thereby increasing its bioavailability within cells.[309] The mechanism by which quercetin acts as a zinc ionophore involves its ability to form complexes with zinc ions. This formation allows quercetin to transport zinc efficiently across lipid membranes, which are otherwise impermeable to charged ions like zinc. Once inside the cell, zinc can participate in various biochemical functions, including acting as a cofactor for numerous enzymes and providing antioxidant protection against oxidative stress. This can enhance the antioxidant effects of zinc and support its role in cellular functions, particularly in combatting oxidative stress.

Elemental zinc is just pure zinc by itself, like a plain metal. You won't find it this way in supplements because our bodies can't easily absorb it in this form. Instead, the zinc in supplements is always combined with another substance, like a carrier molecule (such as bisglycinate, gluconate, citrate, or picolinate). Think of these carriers as escorts that help the zinc get into your body more effectively. The amount of actual, usable zinc (the elemental zinc) in a supplement depends on which carrier it's attached to, so a 50 mg zinc bisglycinate supplement won't have 50 mg of pure zinc. On average, zinc bisglycinate contains 25% elemental zinc by weight, meaning a 50 mg supplement provides 12.5 mg of elemental zinc. The label usually tells you how much elemental zinc you're actually getting.

When it comes to zinc bioavailability, form matters. Some older studies suggested that zinc picolinate might be better absorbed than other forms.[310] Zinc gluconate, frequently used in supplements and lozenges, has demonstrated good bioavailability–60.9% according to one study (range 50.6%–71.7%).[311],[312] Studies have shown that zinc citrate has comparable absorption rates to zinc gluconate and is better absorbed than zinc oxide—61.3% average (range 56.6%–71.0%).[313] It also has favorable sensory properties, making it suitable for syrups and chewable tablets. Some research indicates that zinc bisglycinate may have higher bioavailability compared to zinc gluconate due to its transport through peptide channels, exhibiting 43.4% greater bioavailability.[314],[315] Zinc oxide and zinc sulfate are considered poorly absorbed forms. A laboratory model compared various forms and reported zinc bioavailability as: zinc bisglycinate 5.77%–9.38%, zinc gluconate 4.48%–6.19%, zinc methionine 4.48%, zinc picolinate 3.15%, zinc citrate 1.99%, and zinc sulfate 1.13%.[316] These findings demonstrate that chelated forms (bisglycinate, picolinate, gluconate, methionine, citrate) are more bioavailable than non-chelated forms (sulfate, oxide). It should also be noted that dietary phytates—found in whole grains, legumes, nuts, and seeds—significantly decrease zinc absorption, while dietary protein increases zinc absorption.[317] The current evidence suggests that zinc bisglycinate is the superior form for bioavailability. As an added bonus, it tends to be gentler on the stomach and less likely to cause digestive upset.

A meta-analysis of 10 randomized controlled trials (RCTs) investigated the impact of oral zinc supplementation on serum markers of oxidative stress, including malondialdehyde (MDA), total antioxidant capacity (TAC), glutathione (GSH), and nitric oxide (NO).[318] The dosage of elemental zinc ranged from 11 to 100 mg per day, with gluconate and sulfate being the most common form investigated. Zinc supplementation significantly reduced MDA levels (a marker of lipid peroxidation and oxidative stress) and increased serum TAC and GSH levels (indicators of antioxidant defense). No significant changes were observed in NO levels. A non-linear relationship—meaning increasing the dose doesn't result in a predictable, consistent increase or decrease in MDA levels—was found between zinc dosage and MDA levels. The findings of the meta-analysis encourage the use of zinc for oxidative stress-related conditions.

Another meta-analysis looked at 25 clinical trials assessing the effects of zinc on oxidative stress and inflammatory markers.[319] The amount and forms of zinc consumed daily varied widely from 15 to 240 mg (not elemental, but total supplement) with the most common forms being gluconate and sulfate. Zinc supplementation significantly reduced CRP (an inflammatory marker), IL-6 (another inflammatory cytokine), and MDA (oxidative stress marker). No significant changes were noted in nitric oxide (NO) and the increase in GSH was not statistically significant. Zinc supplementation may have beneficial anti-inflammatory and anti-oxidative effects in adults.

A systematic review and meta-analysis of 23 controlled clinical trials examined the efficacy of zinc supplementation in improving the antioxidant defense system by evaluating GPX, GSH, total antioxidant capacity, and SOD levels..[320] 30 to 150 mg of elemental zinc was administered daily in the trials, which was most commonly gluconate or sulfate. Zinc supplementation significantly increased GSH and TAC levels. After adjusting for publication bias, zinc also showed a significant increasing effect on SOD activity. No significant effect was found on GPX activity. The conclusion of the review was that zinc supplementation can improve certain components of the antioxidant defense system in humans.

A dose-response meta-analysis is a type of meta-analysis that goes beyond just comparing treatment versus no treatment. Instead, it looks at how different doses of a treatment (in this case zinc) affect an outcome—quantifying the relationship between dose and effect across multiple studies. A dose-response meta-analysis of 18 studies evaluated the effect of zinc supplementation on MDA and homocysteine (Hcys) levels, as markers of oxidative stress..[321] Zinc supplementation significantly reduced both MDA and Hcys levels. The maximum reducing effect on MDA was observed with zinc dosages below 40 mg/day, meaning taking more than this doesn't improve the effect. No significant effect was found on another oxidative stress marker, TBARS. Zinc supplementation, particularly at doses under 40 mg/day, may be a beneficial approach to reduce oxidative stress by lowering MDA and Hcys levels.

These meta-analyses and systematic reviews of human clinical trials suggest that zinc supplementation can positively influence several biomarkers of oxidative stress, generally by reducing markers of damage (like MDA) and enhancing antioxidant capacity (like TAC and GSH). These findings support the potential role of zinc in managing conditions where oxidative stress plays a significant role. However, the specific effects and optimal dosages may vary depending on the condition, the individual's zinc status, and other factors. Further research is warranted to clarify these aspects and to explore the clinical benefits of zinc supplementation in various oxidative stress-related diseases.

Zinc can be found in various dietary sources, such as meat, shellfish, legumes, seeds, nuts, dairy products, and whole grains. It exists in several forms, each having different levels of bioavailability. Generally, zinc from animal sources tends to be better absorbed than that from plant sources due to the presence of phytates in plants, which can bind zinc and inhibit its absorption. Therefore, ensuring a balanced diet that includes a variety of zinc-rich foods is crucial for maintaining optimal zinc levels, thus supporting overall health, particularly in combating oxidative stress.

Quercetin

Preferred Form(s): Quercetin phytosome or liposomal quercetin | *Typical Adult Dosage*: Quercetin phytosome—250 mg 1–2 times daily; Liposomal quercetin—250 mg, 1–3 times daily; Quercetin dihydrate, maintenance—500–1,000 mg/day; Quercetin dihydrate, intensive—500 mg, four times daily | *Companion Nutrients*: Vitamin C, vitamin E, zinc, magnesium, bromelain, curcumin, resveratrol, fish oil | *When to Take*: Morning or midday with food

Quercetin is a powerful flavonoid found in many fruits, vegetables, and grains, particularly in foods like apples, onions, berries, and citrus fruits. One of its key benefits is its ability to scavenge free radicals that can cause cellular damage and contribute to oxidative stress. Quercetin is particularly effective against harmful free radicals such as superoxide and hydroxyl radicals, which are known to damage cells and tissues.

In addition to its antioxidant properties, quercetin also has strong anti-inflammatory effects.[322] It helps inhibit the production of inflammatory substances in the body, making it beneficial for those suffering from chronic inflammatory conditions. For individuals with respiratory problems, like asthma, quercetin may reduce inflammation in the airways and improve overall respiratory function,[323, 324] helping to ease symptoms and enhance quality of life.

Quercetin can enhance immune function by supporting various immune cells in fighting off infections and reducing excessive inflammatory responses. By doing so, it plays a role in bolstering overall health, particularly during cold and flu season or in response to allergens. It's also a powerful natural solution for allergic responses, suggesting it has a balancing effect on the immune system.[325] When considering all of its benefits and properties, quercetin is a supplement that could benefit many.

As for its forms and bioavailability, quercetin is available in multiple supplement forms. However, a notable limitation associated with quercetin is its poor solubility in water and consequently low bioavailability, which can hinder its clinical efficacy.[326] Some studies suggest that quercetin can be enhanced when paired with other nutrients, such as vitamin C, which may improve its absorption and effectiveness. Additionally, certain formulations of quercetin, such as those combined with lipid-based carriers or quercetin dihydrate, can further improve its bioavailability, allowing for higher amounts to reach the bloodstream and exert their beneficial effects. Encapsulating quercetin in liposomal delivery systems enhances its bioavailability and ability to scavenge free radicals.[327] Research suggests that liposomal quercetin is eight to nine times more bioavailable than standard quercetin.[328] Forming complexes with substances like methyl-β-cyclodextrin can significantly increase quercetin's water solubility. One study noted a 254-fold increase in water solubility and a 10% enhancement in antioxidant activity with this method.[329] Another form that enhances bioavailability is quercetin phytosome, a delivery system based on food-grade lecithin that has been shown to improve oral absorption. A clinical study found that this form is 20 times more bioavailable than standard quercetin in healthy individuals.[330] Lastly, fenugreek galactomannans as part of a self-emulsifying reversible hybrid-hydrogel seem to improve the oral absorption of quercetin 62-fold according to research.[331] This form offers significant improved solubility, stability, and bioavailability. As of this writing, quercetin phytosome, liposomal quercetin, and quercetin dihydrate are readily commercially available.

Meta-analyses of human clinical trials have demonstrated that quercetin supplementation can lead to significant reductions in both systolic and diastolic blood pressure.[332] Notably, the blood pressure-lowering effect of quercetin appears to be dose-dependent, with studies using dosages of 500 mg per day or higher showing more pronounced reductions. This suggests that a threshold dose might be necessary to produce a clinically meaningful effect on blood pressure. Beyond its impact on blood pressure, quercetin has also been shown to possess anti-atherosclerotic (helping prevent plaque buildup in the arteries), antiproliferative (helping slow abnormal or excessive cell growth), and anti-inflammatory properties,[333] all of which are beneficial in maintaining cardiovascular health. The potential of quercetin to address multiple risk factors associated with cardiovascular disease highlights its therapeutic promise in this area.

A meta-analysis of randomized controlled trials indicated that quercetin supplementation, particularly at a dosage of 1000 mg per day for more than seven days, can promote recovery following intense exercise.[334] This analysis revealed significant decreases in muscle soreness reported within 0 to 24 hours after exercise, as well as a reduction in creatine kinase levels, a key marker of muscle damage, at 24 to 48 hours post-exercise. Additionally, quercetin supplementation was associated with lower levels of post-exercise oxidative stress. However, this particular meta-analysis did not find a significant effect of quercetin on the concentration of interleukin-6 (IL-6), an inflammatory marker, in the context of exercise recovery. In contrast, another study investigating the effects of short-term (seven-day) quercetin supplementation (1,000 mg per day) in healthy adults observed improved post-exercise insulin sensitivity, increased total antioxidant capacity (TAC and SOD activities), and reduced levels of MDA, a marker of lipid peroxidation, during the recovery period.[335] This study also reported enhanced cycling time to exhaustion following quercetin treatment, accompanied by lower levels of IL-6 and creatine kinase at 24 hours post-exercise. These findings suggest that quercetin may be a beneficial ergogenic aid by mitigating exercise-induced oxidative stress and inflammation, thereby aiding muscle recovery and potentially enhancing athletic performance. The slight discrepancies in the findings regarding inflammation across different studies warrant further investigation to fully reveal the role of quercetin in modulating the inflammatory response to exercise.

Quercetin has demonstrated broad anti-inflammatory activities through its ability to modulate various inflammatory pathways and markers. Clinical evidence suggests its potential in managing specific inflammatory conditions. For instance, a study indicated that quercetin supplementation reduces markers of both oxidative stress and inflammation in individuals with sarcoidosis.[336] Furthermore, research has explored the potential of quercetin in alleviating symptoms and reducing inflammatory markers in conditions such as rheumatoid arthritis and diabetic kidney disease.[337,338] These observations highlight the therapeutic potential of quercetin in addressing a range of inflammatory disorders, likely through its multifaceted effects on oxidative stress and inflammatory signaling.

Quercetin has been identified as a potential therapeutic agent for type 2 diabetes due to its antioxidant, anti-inflammatory, hypoglycemic, and hypolipidemic activities.[339,340] Findings from preclinical trials suggest that quercetin supplementation may help reduce postprandial hyperglycemia and improve insulin sensitivity in individuals with this condition.[341] These effects are likely mediated through quercetin's ability to combat oxidative stress and inflammation, which are key factors in the pathogenesis of diabetes and its complications. The evidence from emerging studies supports the potential of quercetin as a supplementary approach in the management of type 2 diabetes, particularly in improving glucose metabolism.

A comprehensive systematic review showed that quercetin supplementation may exert beneficial effects in the context of sepsis, a life-threatening condition characterized by a dysregulated host response to infection.[342] The review suggests that quercetin can attenuate inflammation and oxidative stress, downregulate the expression of toll-like receptors (TLRs), modulate the immune response, and alleviate organ dysfunctions associated with sepsis. While these findings are promising, the authors of the review concluded that future human randomized clinical trials are warranted to further investigate and confirm the therapeutic potential of quercetin in individuals with sepsis. The current evidence provides a rationale for further clinical investigation into the use of quercetin as an adjunct therapy for sepsis.

Research explored the potential relationship between quercetin intake and oxidative stress in individuals experiencing persistent symptoms following COVID-19 infection.[343] The evidence suggests that persistent COVID is associated with redox abnormalities, and quercetin, with its antioxidant properties, may play a role in mitigating these issues. It can neutralize free radicals, inhibit enzymes that promote cell death, and enhance the production of endogenous antioxidants. A combination of quercetin with Vitamin C may also disrupt virus entry and replication while strengthening the immune response. Further research, particularly in the form of human clinical trials, is needed to determine the clinical utility of quercetin in managing long-term COVID symptoms.

Quercetin is a highly beneficial flavonoid that combats oxidative stress and inflammation, offering protective effects against various health conditions. Its bioavailability can be optimized through careful dietary choices and supplementation strategies. By incorporating quercetin-rich foods into the diet, such as apples, onions, and citrus fruits, or considering high-quality supplements, individuals can effectively harness its health benefits and support their overall well-being.

N-Acetylcysteine (NAC)

Preferred Form(s): N-Acetylcysteine ethyl ester (NACET) | *Typical Adult Dosage*: NACET—50–100 mg/day; NAC maintenance—600-1,200 mg/day, NAC intensive—600 mg three to four times daily | *Companion Nutrients*: Glycine, glutamine, alpha-lipoic acid, vitamin C, vitamin E, selenium, zinc, quercetin | *When to Take*: Morning to early afternoon on an empty stomach

N-acetylcysteine (NAC) is a powerful antioxidant and a precursor to glutathione, one of the body's most important natural antioxidants. By increasing the levels of glutathione in the body, NAC plays a crucial role in combating oxidative stress and is therefore a promising solution for various oxidative stress-related conditions, including chronic diseases, neurodegenerative disorders, respiratory issues, and cardiovascular problems. NAC works by neutralizing harmful free radicals and supporting the body's detoxification process, making it especially beneficial for cellular health and overall well-being.

In addition to its antioxidant properties, NAC has been found to have anti-inflammatory effects. It can help reduce inflammation in the body, which is a critical factor in conditions like asthma, chronic obstructive pulmonary disease (COPD), and other respiratory illnesses.[344] By helping to clear mucus from the airways, NAC can improve respiratory function and ease breathing difficulties, making it a popular supplement among individuals with lung conditions.

A significant limitation of the use of NAC is its poor bioavailability—ranging from 4%–10% in humans.[345] As with other supplements, certain forms can exhibit higher bioavailability, making it more effective. This is the case with n-Acetylcysteine, which is available in formulations combined with glycine or as an ethyl ester form (N-Acetyl-L-Cysteine Ethyl Ester; NACET). NACET has significantly higher bioavailability compared to standard NAC, primarily due to its greater ability to dissolve in fats, oils, and lipids (lipophilicity), which allows it to more easily cross cell membranes.[346,347] Indeed, preclinical research suggests it's bioavailability is 12.2-fold better than standard NAC.[348] NACET rapidly enters cells where is it transformed into NAC and cysteine, significantly increasing glutathione in tissues.[349] It can also cross the blood-brain barrier.[350] Results of research suggests that NACET can improve glutathione levels by over 300% compared to roughly 10% for standard NAC, a massive 30-fold increase.[351] Based on this, it also has superior effectiveness against oxidative stress.

A novel form of NAC combined with glycine (GlyNAC) shows improved bioavailability compared to standard NAC.[352] Research demonstrated that supplementation with GlyNAC in older adults improved GSH deficiency, oxidative stress, mitochondrial dysfunction, and inflammation.[353,354,355] The study found significant improvements in insulin resistance, endothelial function, muscle strength, and cognitive function, indicating that GlyNAC could be a promising intervention for age-related oxidative stress and associated conditions. You can also find supplements that contain glycine and NACET (Gly-NACET), which may further enhance bioavailability and effectiveness.

NAC's ability to cross cell membranes effectively supports its role in protecting various tissues from oxidative damage, including the liver and brain. This feature is particularly important for its use as a treatment for acetaminophen (paracetamol) overdose, where it helps replenish glutathione levels that become depleted in such toxic situations.

A meta-analysis of 28 randomized controlled trials investigated the effects of NAC on inflammatory and oxidative stress biomarkers.[356] Dosage used in the studies were between 500–2,000 mg/day, which was taken from nine to 180 days. The study found that NAC significantly decreased malondialdehyde (MDA) (a marker of oxidative stress) and interleukin-8 (IL-8) levels. Sensitivity analysis also showed significant effects of NAC in reducing TNF-alpha and IL-6, and homocysteine levels. No significant change was observed in C-reactive protein (CRP) levels.

A review conducted by the Nutritional Medicine Institute highlighted NAC's direct antioxidant and anti-inflammatory effects.[357] It notes a 2020 meta-analysis showing a significant decrease in MDA with NAC supplementation (400-2000 mg/day) and several studies demonstrating increased glutathione (GSH) levels in blood cells following NAC administration. The review also discusses NAC's role in modulating inflammation by inhibiting NF-κB and pro-inflammatory cytokines. The researchers determined that "there is evidence for benefits of NAC in diverse conditions, including respiratory infections, various mental health disorders, male infertility and polycystic ovary syndrome.

A meta-analysis of 13 studies (4,155 people with COPD) found that NAC treatment was associated with significantly fewer exacerbations of chronic bronchitis or COPD.[358] This effect was more apparent in patients without evidence of airway obstruction, but high doses of NAC (≥1200 mg/day) were also effective in people with COPD diagnosed by spirometry. Individuals without airway obstruction but experiencing chronic bronchitis responded well to 600 mg/day. NAC was well-tolerated. A contradictory meta-analysis of nine randomized controlled trials concluded that long-term oral NAC therapy did not reduce the risk of acute exacerbation of COPD symptoms, including reduction of the decline in lung volume in people with COPD.[359] However, it should be noted that all but two of the trials limited dosing to only 600 mg/day, which may be lower than the therapeutic dose necessary to produce meaningful improvements. Further expounding on this conclusion, the study authors noted research showing benefits for NAC supplementation for COPD, respiratory infections, and COVID-19. Another meta-analysis confirmed that NAC could improve the symptom improvement rate of people with acute exacerbation of COPD (AECOPD), improve lung function (FEV1), and enhance the body's antioxidant capacity.[360]

Many mental health conditions are associated with oxidative stress, reduced antioxidant capacity, and inflammation, so it's unsurprising that studies show NAC is beneficial for a variety of mental health conditions. This is due to its antioxidant, anti-inflammatory, and neuromodulatory effects. A review and meta-analysis found that adjunctive NAC significantly improved total psychopathology in schizophrenia but had no significant effect on depressive and manic symptoms in bipolar disorder and only a small effect on major depressive symptoms.[361] NAC was found to be a safe treatment. A more recent review suggests NAC's potential as an adjunctive therapy for various neuropsychiatric disorders due to its modulation of neurotransmitter systems, oxidative balance, and inflammatory mediators.[362] It highlights promising results in bipolar disorder, depression, OCD, anxiety disorders, and schizophrenia, warranting further investigation. Most trials in the review used a dosage between 1,200 and 2,400 mg daily. Findings of one review determined that NAC supplementation increases GSH brain levels and balances neurotransmitters, in particular glutamate homeostasis.[363,364] Research using NAC for addictions, ADHD, autism, depression, obsessive compulsive disorder, schizophrenia, and bipolar disorder were analyzed. NAC was administered at doses of 1200 to 3,600 mg/day for four to 25 weeks during the trials included in the review. The researchers concluded that NAC shows promising results in people with these disorders, including among individuals in whom previous treatment was not fully effective. While the research is mixed on mental health and NAC, it appears to be a promising adjunct solution.

Men experiencing infertility often have elevated levels of ROS in their semen, which almost acts like rust for sperm. When there's an excess of these rust molecules, it can lead to several problems for sperm. This includes sperm dysfunction, where sperm struggle to move or fertilize an egg effectively. Additionally, these high ROS levels can cause sperm DNA damage, harming the genetic material crucial for a healthy pregnancy. Ultimately, this damage significantly reduces a man's reproductive potential, making it more difficult for him to father a child. A clinical trial evaluated the use of NAC among men experiencing infertility.[365] The research demonstrated that supplementing with 600 mg of NAC for three to six months increased male fertility by improving semen quality and promoting higher testosterone levels.

Oxidative stress is strongly linked to Polycystic Ovary Syndrome (PCOS). In women with PCOS, there's often an excess of ROS and a reduced antioxidant defense, leading to cellular damage. This oxidative stress is deeply intertwined with glucose metabolism, particularly insulin resistance. Insulin resistance, a hallmark of PCOS in which cells don't respond well to insulin, leads to higher insulin levels in the blood (hyperinsulinemia). This elevated insulin can further increase ROS production, creating a vicious cycle that worsens oxidative stress. This increased oxidative stress then contributes to the inflammation and impaired glucose regulation seen in PCOS, potentially harming ovarian function and contributing to other metabolic complications like a higher risk of type 2 diabetes. A systemic review and meta-analysis evaluated evidence for the use of NAC for PCOS.[366] The specifics for polycystic ovary syndrome (PCOS) usually involved taking 1,200–1,800 mg of NAC daily, which promoted better ovulation, fertility, sex hormone levels, and oocyte quality.

Another area of conflicting research involves liver disease. A meta-analysis from 2021 found no significant difference in overall survival between NAC and placebo in adults with acute non-acetaminophen, non-alcoholic, and non-viral hepatitis.[367] There was also no difference in infection rates. Contrarily, a 2022 review and meta-analysis of prospective studies on non-acetaminophen-induced acute liver failure found that NAC was associated with significantly higher odds of overall survival, transplant-free survival, and post-transplant survival, as well as a shorter hospital stay.[368] Many of these studies used intravenous NAC at relatively large doses. A systematic review and meta-analysis of eight eligible studies assessing the effects of NAC on liver function found no significant changes in liver enzymes (AST, ALT, ALP), but notable improvements in albumin and bilirubin levels.[369] These summaries highlight the ongoing research into NAC's effects on various conditions linked to oxidative stress, with some promising results in specific areas, while others require further investigation with well-designed, large-scale clinical trials.

One important consideration with NAC is that some individuals are sensitive to sulfur-containing compounds, which may lead to adverse effects. For example, NAC can interfere with sleep in these individuals if taken at night. One way to overcome this sensitivity is to ensure NAC is taken with molybdenum. Molybdenum is an essential cofactor for the enzyme sulfite oxidase (SO). SO is crucial for the breakdown of sulfites (a common byproduct of sulfur metabolism and found in many foods and medications) into harmless sulfates, which are then excreted. Since NAC is a precursor to cysteine, its metabolism will involve pathways that produce sulfites. If insufficient levels of molybdenum are present, SO activity will be diminished, leading to the accumulation of sulfites in the body, which are toxic, especially to the brain and nervous system. Symptoms of sulfite sensitivity or sulfite toxicity due to impaired SO activity can include headaches, fatigue, brain fog, hives, wheezing, asthma, digestive issues (diarrhea), skin problems, rapid heart rate, and even severe allergic-like reactions.

Think of it this way: your body has a special clean-up crew that handles sulfur from food and supplements like NAC. Molybdenum is like a key ingredient that helps this crew do its job. If you don't have enough molybdenum, this clean-up crew can't work properly, causing sulfur waste to build up. This buildup can make you feel unwell, leading to symptoms like headaches or upset stomach when you consume sulfur-rich ingredients. So, the bottom line is this: if people are sensitive to sulfites or experience symptoms when taking sulfur-containing supplements like NAC, adding 400–500 mcg of molybdenum can help the body process sulfur more efficiently and reduce those unwanted reactions.

NAC is a potent antioxidant that helps combat oxidative stress by boosting glutathione production and reducing inflammation. Its various forms, along with its multiple therapeutic actions, make it an accessible and valuable supplement for promoting overall health and supporting the body's natural defenses against oxidative damage.

Astaxanthin

Preferred Form(s): Astaxanthin oil/softgel, liposomal or water-soluble/dispersible forms | *Typical Adult Dosage*: 12–24 mg/day | *Companion Nutrients*: Fish oil, lutein, zeaxanthin, beta-carotene, vitamin E (specifically tocotrienols), CoQ10, choline, curcumin | *When to Take*: Anytime during the day with food

Astaxanthin is a naturally occurring carotenoid, which is a type of pigment found in various plants and algae. It is most commonly derived from microalgae, particularly *Haematococcus pluvialis*, and is also found in seafood like salmon, shrimp, and trout, which acquire it from eating the microalgae. One of the most remarkable characteristics of astaxanthin is its powerful antioxidant properties. Unlike some other antioxidants, astaxanthin is particularly effective at neutralizing singlet oxygen—an excited, high-energy form of oxygen that can cause significant cellular damage—as well as a variety of free radicals. This ability to scavenge a wide range of reactive oxygen species makes astaxanthin highly effective in combating oxidative stress.

The beneficial effects of astaxanthin extend to multiple areas of health. For eye health, research has shown that astaxanthin can help protect the retina and reduce the risk of age-related macular degeneration by lowering oxidative damage in the eyes.[370] Astaxanthin also offers protection for the skin against ultraviolet (UV) damage from sun exposure, which can lead to premature aging and skin cancer.[371] It helps improve skin elasticity and hydration while reducing redness and inflammation caused by UV exposure.

Furthermore, astaxanthin has cardiovascular benefits as well. It has been shown to improve lipid profiles by reducing oxidative stress on blood vessels and supporting endothelial function, which can lead to healthier blood circulation and a reduced risk of cardiovascular diseases.[372] Its anti-inflammatory properties also make it helpful in addressing conditions related to chronic inflammation, such as arthritis,[373] metabolic syndrome, and other inflammatory diseases.

In terms of how it's taken, astaxanthin is most often available as a softgel or tablet. Because it is fat-soluble, its absorption is best facilitated when taken with dietary fats,[374] which can significantly enhance its bioavailability. Many supplements also include lipids or fats to promote absorption. The bioavailability can vary depending on the source and formulation, so it is important to choose high-quality astaxanthin supplements that are well-absorbed to maximize its health benefits. In this case, a softgel with a fatty oil (e.g., coconut oil, MCT oil, olive oil) may be best.

The bioavailability of astaxanthin is generally lower due to its lipophilic (fat-soluble) nature—estimated between 10% to 50%.[375] Like other supplements, researchers have leveraged advanced delivery systems in attempts to improve the bioavailability of astaxanthin. Preclinical models suggest that liposomal astaxanthin is more bioaccessible because it is more soluble and stable.[376,377] Astaxanthin can also be micronized or nano-sized (reducing its particle size) to improve bioavailability.[378,379] Doing this can increase its surface area and improve dissolution and absorption. Some newer forms use an emulsification process to make astaxanthin more water-soluble, which can also enhance bioavailability.[380] A clinical trial compared various forms of astaxanthin—standard astaxanthin/oil softgels, enteric-coated softgels, water-soluble astaxanthin emulsion, water-dispersible astaxanthin, and liposomal astaxanthin.[381] Surprisingly, there was not a massive difference between the tested formulas with the water-soluble/dispersible and liposomal astaxanthin being only two to three times more bioavailable. If price is similar, the liposomal or water-soluble/dispersible forms are worth using for their greater tissue distribution and potential efficacy.

It's also important to note that natural astaxanthin from microalgae comes in esterified forms (attached to fatty acids), which requires the body to de-esterify it before absorption. Contrarily, synthetic forms or those grown on yeast don't usually have fatty acids attached, which makes some people believe it is more bioavailable. However, research with other carotenoids suggests that bioavailability seems to be more dependent on the formulation (e.g., with lipids) than whether it's esterified or not.[382,383]

A meta-analysis conducted in 2021 specifically investigated the anti-aging effects of astaxanthin in humans, with a primary focus on skin aging.[384] Their comprehensive search of major scientific databases identified eleven relevant studies, nine of which were randomized, controlled human trials assessing the impact of oral astaxanthin supplementation on various skin aging parameters. The pooled analysis of these RCTs demonstrated that oral astaxanthin significantly improved skin moisture content and elasticity compared to a placebo. Specifically, there was a statistically significant difference in moisture content and elasticity. While these findings suggest a positive impact on skin health, the meta-analysis did not find a significant reduction in wrinkle depth with oral astaxanthin supplementation. Two open-label, prospective studies included in the review hinted at slight protective effects of topical and combined oral-topical astaxanthin applications on skin aging. The observed improvements in skin moisture and elasticity, while focused on a specific outcome, can be interpreted as indirect evidence of reduced oxidative damage, since oxidative stress is a known contributor to the degradation of collagen and elastin, which are essential for skin hydration and firmness.

Another comprehensive meta-analysis evaluated the effects of astaxanthin supplementation on a range of oxidative stress and inflammation biomarkers in humans.[385] This analysis included twelve randomized controlled trials encompassing 380 participants who received astaxanthin for at least one week and were compared to a placebo group. A key finding was the significant reduction in blood malondialdehyde (MDA) concentration in the astaxanthin-supplemented group, indicating a decrease in lipid peroxidation, a primary marker of oxidative stress. Notably, this effect was particularly pronounced and statistically significant in people with type 2 diabetes—a condition known for highly elevated oxidative stress. The meta-analysis also suggested potential improvements in SOD activity and a reduction in serum isoprostane levels in overweight subjects, although the number of trials for these specific biomarkers was limited. Furthermore, astaxanthin supplementation significantly reduced blood IL-6 concentration in diabetic individuals, suggesting an anti-inflammatory effect in this population. However, no significant effects were observed on other inflammatory markers such as CRP and TNF-α. The specific reduction of MDA in diabetic individuals suggests a potential benefit of astaxanthin in mitigating the elevated oxidative stress characteristic of this metabolic disorder.

In a systematic review and meta-analysis published in 2020, the antioxidant activity of astaxanthin in humans was assessed based on nine randomized controlled trials published up to 2015.[386] The overall meta-analysis revealed a borderline significant antioxidant effect of astaxanthin compared to the control group, specifically an MDA-lowering effect. However, the authors noted that the data were insufficient for definitive conclusions. When comparing oxidative stress markers within the intervention groups to their baseline levels, astaxanthin supplementation significantly decreased plasma MDA and isoprostane. Subgroup analysis indicated that a high dose of astaxanthin ($\geq$ 20 mg/day) exerted a significant antioxidant effect on total antioxidant capacity, isoprostane, and SOD. Conversely, a low dose (< 20 mg/day) did not show a significant effect. Furthermore, an antioxidant effect was observed after a three-week intervention but not after 12 weeks or three months for isoprostane and SOD. The authors concluded that the antioxidant effect of astaxanthin on humans remained unclear due to the limitations of the available data at the time of their review. The observation that higher doses might be required for significant antioxidant effects underscores the importance of dosage in future research.

A review article published in 2018 provided a broad overview of astaxanthin's biological effects, emphasizing its potent antioxidant and anti-inflammatory capabilities.[387] The authors highlighted astaxanthin's superior free radical inhibitory activity compared to other carotenoids and its ability to lower harmful oxygen molecules, reduce swelling and scarring, and prevent cell death. They also emphasized astaxanthin's influence on cellular signaling pathways, particularly its role in controlling the redox-sensitive transcription factors Nrf2 and NF-κB, which are involved in oxidative stress and inflammation, respectively. This means astaxanthin doesn't just neutralize harmful molecules directly—it also helps turn on the body's own built-in defense and repair systems, making cells better at protecting themselves from ongoing damage.

A meta-analysis from 2021 evaluated the impact of astaxanthin supplementation on blood pressure in adults, including ten RCTs with 493 participants.[388] The analysis showed a marginal decline in diastolic blood pressure (DBP) but not systolic blood pressure (SBP) with astaxanthin supplementation. In participants, astaxanthin significantly reduced DBP by 2.77 mmHg. Dosages of 12 mg/day or more resulted in a slight decrease in DBP, and a significant DBP reduction was observed in participants from Asian countries. This suggests a potential benefit in managing hypertension, particularly diastolic blood pressure, in certain populations.

Another review discussed the antioxidant and anti-inflammatory mechanisms of astaxanthin in cardiovascular diseases.[389] The authors highlighted its ability to modulate lipid and glucose metabolism and prevent disorders like atherosclerosis and high blood pressure. They also mentioned its potential to promote reverse cholesterol transport and improve lipid profiles in individuals with high cholesterol. These findings collectively suggest that astaxanthin may offer multifaceted benefits for cardiovascular health by addressing several underlying risk factors.

A randomized controlled trial investigated the effects of 12 mg/day astaxanthin supplementation for eight weeks in 50 patients with coronary artery disease (CAD).[390] The study found a significant reduction in total cholesterol and LDL-C levels in the astaxanthin group compared to placebo. However, no significant effects were observed on triglycerides, HDL-C, glycemic parameters, anthropometric indices (BMI, waist circumference, waist-to-hip ratio, waist-to-height ratio, body fat percentage), body composition, Sirtuin1 (longevity), or TNF-α (inflammation) levels. This suggests a potential role for astaxanthin in managing high cholesterol, a risk factor for cardiovascular disease when metabolic dysregulation exists simultaneously.

A recent critical review explored the effects of astaxanthin on cognitive function and neurodegeneration in humans.[391] The review highlighted potential benefits in improving cognitive function, facilitating neuroprotection, and slowing neurodegeneration. Studies discussed showed improvements in word recall in younger individuals and some improvements in memory and processing speed in elderly subjects, though they were not always statistically significant. The review also mentioned potential benefits in Parkinson's and Alzheimer's disease models, but human evidence remains limited. Notably, astaxanthin showed more consistent cognitive benefits when combined with other ingredients (sesamin), particularly in improving processing speed and memory. While the evidence for standalone effects is still emerging, the potential, especially in combination therapies, warrants further investigation.

A review published in 2020 suggested beneficial pleiotropic effects—when a single gene influences multiple, seemingly unrelated traits—of astaxanthin in the prevention and treatment of several ocular diseases due to its anti-inflammatory and antioxidant activities.[392] The review called for further human intervention studies to define optimal application modalities. A quasi-experimental study evaluated the effect of 6 mg astaxanthin twice daily for 30 days in individuals with mild-to-moderate dry eye disease (DED).[393] The study found significant improvements in OSDI score, tear break-up time, corneal fluorescein staining, and other markers of DED, suggesting a specific clinical application for astaxanthin in managing this condition. An *in vitro* study compared lutein and astaxanthin's protective properties against blue-violet light on corneal epithelial cells.[394] Both compounds showed protective effects against phototoxicity and ROS production, with lutein being slightly more efficient in this model. These findings suggest that astaxanthin may contribute to eye health by mitigating oxidative stress and inflammation.

There is a strong agreement in the literature regarding the potent antioxidant properties of astaxanthin in humans. Additionally, consistent evidence supports the benefits of oral astaxanthin supplementation for improving skin moisture and elasticity. An emerging consensus also points to the potential benefits of astaxanthin in managing oxidative stress and inflammation in patients with type

2 diabetes, and promising evidence suggests its role in improving symptoms of dry eye disease.

In summary, astaxanthin is a potent carotenoid antioxidant that plays a significant role in protecting against oxidative stress-related damage. It supports eye health, skin protection from UV radiation, and cardiovascular health while also offering anti-inflammatory benefits. Its availability in various supplement forms and its fat-soluble nature can help maximize absorption, making it a valuable addition to a comprehensive plan to build resilience against oxidative stress.

Vitamin A | Beta-Carotene

Preferred Form(s): Vitamin A (retinol) | *Typical Adult Dosage*: Maintenance—10,000 IU/day; Intensive—25,000 IU/day | *Companion Nutrients*: Vitamin E, vitamin C, B vitamins, beta-carotene, zinc, fish oil | *When to Take*: Anytime during the day with food

Vitamin A is an essential fat-soluble vitamin that comes in two primary forms: preformed vitamin A (also known as retinol) and provitamin A carotenoids, such as beta-carotene. Retinol is found in animal sources like liver, fish, and dairy products, while beta-carotene, which the body can convert into retinol, is abundant in colorful fruits and vegetables, particularly carrots, sweet potatoes, spinach, and kale. As a vital antioxidant, vitamin A plays a significant role in combatting oxidative stress by effectively neutralizing ROS, including harmful singlet oxygen and free radicals that can cause damage to cells and tissues.

The antioxidant properties of vitamin A are crucial for several aspects of health. One of its most well-known roles is maintaining healthy vision.[395] Vitamin A is a key component of rhodopsin, a pigment in the retina that is essential for low-light and color vision. By protecting eye tissues from oxidative damage, vitamin A can help prevent conditions such as age-related macular degeneration (AMD) and cataracts, both of which are associated with the aging process and oxidative stress. In addition to vision, vitamin A also supports skin integrity, promoting cellular regeneration and helping to combat the signs of aging.[396] Its antioxidant effects play a role in protecting the skin from environmental damage, including UV radiation and pollution.

Moreover, vitamin A is vital for maintaining a robust immune system.[397] It helps regulate immune responses and is involved in the production of white blood cells, which are essential for fighting infections. Adequate levels of vitamin A can reduce the risk of infections and chronic diseases that are linked to oxidative stress and inflammation. This underscores the importance of maintaining optimal vitamin A levels for overall health and well-being.

When it comes to bioavailability, the absorption of vitamin A varies between its two forms. Preformed vitamin A is readily absorbed and utilized by the body (absorption rates ranging from 70%–90%),[398] while beta-carotene must first be converted into retinol, which can be influenced by several factors, including an individual's genetic makeup and dietary fat intake. This conversion process is not always efficient and can vary greatly between individuals (absorption ranges from 8.7%–65%).[399] Therefore, preformed vitamin A is generally considered more bioavailable from supplements. Consuming beta-carotene with healthy fats can enhance its absorption, making it more effective. Some individuals, such as those with gastrointestinal disorders, may have difficulty absorbing fat-soluble vitamins, and thus may benefit from supplementation.

A meta-analysis evaluated six randomized placebo-controlled trials on the relationship between a plant-based multivitamin/mineral supplement (containing vitamin A as beta-carotene) and oxidative stress.[400] While individual studies had inconsistent results, the overall meta-analysis suggested a general benefit of the supplement in alleviating oxidative stress by providing exogenous oxidant scavengers. However, the effect specific to vitamin A or its precursor was not isolated.

A review of preclinical and clinical studies looked at the effect of beta-carotene, tocopherols, and ascorbic acid on oxidative stress biomarkers.[401] The researchers concluded that all three vitamins are valuable to mitigate oxidative stress in various biological systems. The review highlighted that beta-carotene acts as an antioxidant in two main ways: it can absorb the energy from one type of damaging oxygen molecule, or it can directly react with other harmful free radicals to neutralize them. Beta-carotene can also help get rid of other types of free radicals like superoxide, hydroxyl radicals, and hydrogen peroxide, often changing its own structure in the process.

Emerging research suggests that dietary intake of vitamin A and beta-carotene can influence mood. A meta-analysis of 25 observational studies (100,955 participants) found that dietary vitamin A or beta-carotene was inversely associated with depression.[402] Depression is increasingly understood to be linked to oxidative stress.[403] This imbalance of free radicals and antioxidants can damage cells, including brain cells, and disrupt normal brain function, contributing to the development and progression of depression. Several processes in depression, like the stress response and inflammation in the brain, can increase the production of these damaging free radicals. This research suggest that vitamin A and beta-carotene may protect brain cells from oxidative damage and potentially support healthier brain function.

In summary, vitamin A is an essential nutrient with powerful antioxidant properties that help neutralize oxidative stress, particularly benefiting eye health, skin integrity, and immune function. The presence of both preformed vitamin A and provitamin A carotenoids in a balanced diet is crucial for maintaining optimal health, as it helps to shield the body from oxidative damage and supports various physiological functions necessary for overall well-being.

Pycnogenol (French Maritime Pine Bark Extract)

Preferred Form(s): French maritime pine bark extract standardized to 65%–95% proanthocyanidins | *Typical Adult Dosage*: 100–450 mg/day | *Companion Nutrients*: Vitamin C, zinc, L-arginine, gotu kola | *When to Take*: Morning or midday with or without food

French Maritime pine bark (FMPB), or Pycnogenol®, is a natural plant extract derived from the bark of the French maritime pine tree, scientifically known as *Pinus pinaster* (syn. *P. maritima*). It is renowned for its rich content of proanthocyanidins, a group of powerful antioxidants that exhibit strong protective effects against free radicals, including hydrogen peroxide and superoxide. As previously noted, free radicals can cause oxidative stress, leading to cellular damage and various health issues. By neutralizing these harmful molecules, FMPB extract helps to reduce oxidative damage in the body, contributing to overall health and well-being.

One of the prominent benefits of FMPB extract is its support for cardiovascular health.[404] Research has shown that FMPB extract can improve blood flow by promoting the production of nitric oxide, a molecule that helps blood vessels relax and widen. This enhancement in blood circulation can lead to better oxygen delivery to tissues and organs, as well as lower blood pressure. Additionally, FMPB extract has anti-inflammatory properties that can further benefit heart health by reducing inflammation in blood vessels, which is a contributing factor in many cardiovascular diseases.[405]

Beyond cardiovascular benefits, FMPB extract is also recognized for its positive effects on skin health.[406] It combats oxidative stress related to aging and environmental factors, such as UV exposure and pollution, which can lead to premature skin aging and various skin disorders. By supporting the skin's resilience and aiding in its regeneration, FMPB extract may help improve skin elasticity, hydration, and overall appearance. This makes it a popular ingredient in skincare products and dietary supplements aimed at promoting youthful skin.

When it comes to its forms and bioavailability, FMPB extract is considered highly bioavailable.[407] It can be found in topical formulations for skin health as well. The bioavailability of FMPB extract can be quite favorable due to its natural origin and the way it is absorbed in the digestive tract, although absorption may be enhanced when taken with food containing fats.

The scientific literature reflects extensive investigation into the potential of Pycnogenol, with over 100 research trials conducted to date. This substantial body of work includes numerous reviews and meta-analyses that have synthesized the findings from human clinical trials, providing valuable insights into its efficacy across various health domains. A significant portion of this research has focused on rigorous "gold standard" randomized, double-blind, placebo-controlled (RDP) trials, which are crucial for establishing the true efficacy of an intervention.

According to a review of RDPs, several clinical studies have shown that Pycnogenol, a trademarked FMPB extract, helps reduce oxidative stress in the body through multiple mechanisms.[408] In both short- and long-term studies, doses ranging from 100 to 200 mg per day for periods of two weeks up to six months significantly lowered markers of oxidative damage. For example, in individuals with coronary artery disease, 200 mg/day for eight weeks led to a significant reduction in 15-F(2t)-Isoprostane, a marker of lipid peroxidation. Another study in elderly subjects showed that 150 mg/day for three months reduced plasma F2-isoprostane levels by nearly 23%. The antioxidant effects of Pycnogenol work by neutralizing free radicals, boosting the body's antioxidant enzymes like glutathione, and reducing pro-inflammatory markers like NF-κB and COX enzymes. These actions help protect cells from damage and support better vascular and cognitive health. Overall, Pycnogenol is a well-tolerated and natural way to enhance the body's defense against oxidative stress.

Cardiovascular health represents a significant area of focus in Pycnogenol research, with numerous reviews and meta-analyses examining its effects.[409,410,411] Endothelial function, the ability of blood vessels to dilate properly, is crucial for preventing atherosclerosis and subsequent cardiovascular events. Studies have indicated that Pycnogenol exerts beneficial effects on endothelial function. Notably, a randomized, double-blind, placebo-controlled, cross-over study involving patients with coronary artery disease (CAD) demonstrated that Pycnogenol treatment (200 mg/day for eight weeks) was associated with a significant improvement in flow-mediated dilatation (FMD), a measure of endothelial function. This improvement was linked to a reduction in 15-F(2t)-Isoprostane, a biomarker of oxidative stress.

The impact of Pycnogenol on blood pressure has also been extensively studied. Meta-analyses have suggested that Pycnogenol supplementation leads to significant reductions in both systolic and diastolic blood pressure.[412],[413],[414] These effects appear to be more pronounced in individuals with hypertension (systolic blood pressure $\geq$ 140 mmHg or diastolic blood pressure $\geq$ 90 mmHg) and in trials with longer intervention durations, exceeding 12 weeks. One meta-analysis of 27 randomized controlled trials found that Pycnogenol significantly reduced both systolic and diastolic blood pressure, with subgroup analyses showing consistent systolic blood pressure lowering with both short-term and long-term supplementation at specific dosage ranges.[415]

Regarding lipid profiles, some research suggests potential benefits of Pycnogenol. Meta-analyses have indicated that Pycnogenol supplementation may lead to reductions in LDL and total cholesterol levels.[416],[417] Furthermore, improvements in microcirculation have been observed in patients with coronary artery disease following Pycnogenol treatment. The consistent findings across multiple studies and meta-analyses suggest that Pycnogenol may play a multifaceted positive role in cardiovascular health by targeting endothelial function, blood pressure regulation, and potentially influencing lipid metabolism. The observation that longer intervention periods and specific dosages might yield more pronounced effects on certain cardiovascular outcomes suggests a potential dose-response relationship and highlights the need for further long-term studies to optimize its therapeutic use.

The influence of Pycnogenol on cognitive function has been explored in numerous reviews and studies. Evidence suggests that Pycnogenol may improve various aspects of cognitive performance, including attention, memory, and executive functions in both healthy adults and the elderly.[418],[419],[420] For instance, studies have demonstrated improvements in working memory and a reduction in lipid peroxidation products in elderly individuals taking Pycnogenol. Research involving healthy professionals with increased oxidative stress showed that Pycnogenol supplementation (150 mg/day for 12 weeks) led to significant improvements in spatial recognition memory and mood parameters such as alertness and contentedness.[421] Another study in healthy subjects aged 55-70 with high oxidative stress found that 12-month supplementation with Pycnogenol (100 mg/day) improved cognitive function and daily tasks, and significantly decreased oxidative stress levels.[422] The observed cognitive benefits of Pycnogenol are likely attributable to its antioxidant and vascular effects. By improving blood flow and reducing oxidative damage in the brain, Pycnogenol may protect neuronal cells and support cognitive function.

Diabetes and its associated complications, particularly diabetic retinopathy, have been the subject of research on Pycnogenol. Meta-analyses have indicated that Pycnogenol supplementation may improve blood glucose control, with significant reductions observed in fasting blood glucose and HbA1c levels.[423,424,425,426] The antioxidant properties of Pycnogenol are particularly relevant in the context of diabetes, as oxidative stress plays a crucial role in the pathogenesis of the disease and its complications.

The effects of Pycnogenol on asthma, another condition associated with inflammation and oxidative stress, has been examined in several reviews and studies. Research suggests that Pycnogenol may improve asthma symptoms and lung function. A randomized, placebo-controlled, double-blind study involving children aged six to 18 with mild-to-moderate asthma demonstrated that the group receiving Pycnogenol (1 mg/lb BW or 100 mg daily) experienced significantly greater improvements in pulmonary functions and asthma symptoms compared to the placebo group.[427] Additionally, the Pycnogenol group showed a greater ability to reduce or discontinue the use of rescue inhalers, and a significant reduction in urinary leukotrienes, which are chemical messengers that drive inflammation and airway tightening.. Another study evaluating the efficacy of Pycnogenol over a six-month period in patients with stable, controlled allergic asthma found that Pycnogenol administration (100 mg daily) led to better control of asthma signs and symptoms and reduced the need for medication, including inhaled corticosteroids.[428] These findings indicate that the anti-inflammatory and antioxidant properties of Pycnogenol may offer clinical benefits in the management of asthma, potentially serving as a valuable adjunct therapy.

Osteoarthritis, a degenerative joint disease characterized by inflammation and oxidative stress, has also been investigated in relation to Pycnogenol. Reviews suggest that Pycnogenol may reduce joint discomfort, stiffness, and improve physical function in adults with knee osteoarthritis.[429,430] A meta-analysis of nutraceutical supplements for osteoarthritis included one study that specifically evaluated the effect of Pycnogenol (150 mg once daily for 90 days) on patients with knee osteoarthritis. The results of this study, while part of a larger analysis, suggested potential benefits in pain alleviation.[431] While the overall evidence for Pycnogenol in osteoarthritis may be less definitive compared to other conditions, the indication of pain relief in some studies warrants further investigation, particularly given the role of oxidative stress in the pathophysiology of osteoarthritis. Larger, well-designed clinical trials are needed to fully elucidate the potential benefits of Pycnogenol in managing osteoarthritis symptoms.

The evidence from reviews and meta-analyses of human clinical trials suggests that Pycnogenol, a standardized extract from French maritime pine bark, exhibits promising effects against several conditions associated with oxidative stress. Overall, FMPB extract is a powerful antioxidant supplement that plays a multifaceted role in protecting the body from oxidative stress. With its benefits for cardiovascular health, skin integrity, and overall cellular protection, FMPB extract serves as a valuable addition to a health-conscious lifestyle, offering protection against the detrimental effects of free radicals and supporting the body's natural healing processes.

Milk Thistle (Silymarin)

Preferred Form(s): Silybin-phosphatidylcholine complex (phytosome) | *Typical Adult Dosage*: Phytosome—180–360 mg/day; Milk thistle extract standardized to 70%–80% silymarin—150–600 mg, up to three times daily | *Companion Nutrients*: Vitamin E, turmeric. Dandelion root, artichoke extract | *When to Take*: Anytime during the day 30 minutes prior to a meal

Silymarin is the active component derived from the seeds of the milk thistle plant (*Silybum marianum*), and it is well-regarded for its powerful antioxidant properties. This complex formulation consists mainly of flavonolignans, including silibinin (also known as silybin), isosilibinin, silychristin, and silydianin. Together, they exhibit significant activity against free radicals, including hydrogen peroxide and lipid peroxyl radicals. Among these compounds, silibinin is the most active and abundant, and is credited for the bulk of silymarin's beneficial effects. Silibinin itself exists as two diastereomers, silybin A and silybin B, which share similar antioxidant properties but may exhibit slight differences in bioactivity.

One of the most prominent roles of silymarin, particularly silibinin, is its protective effects on liver health.[432] Silymarin supports the liver in detoxifying harmful substances, thereby enhancing its protective mechanisms against damage from alcohol, drugs, and environmental toxins. Research suggests that silymarin can aid in the prevention and treatment of various liver diseases, including cirrhosis, fatty liver disease, and hepatitis.[433] By reducing oxidative stress in liver cells, silymarin helps protect against cell injury and inflammation, promoting overall liver function and health.

In addition to its liver-protective benefits, silymarin extends its antioxidant effects to other tissues, including the skin.[434] Evidence shows that it protects skin cells from oxidative stress caused by UV radiation and pollution, which can lead to premature aging, skin damage, and even cancer.[435] By combating these harmful effects, silymarin helps maintain skin integrity and promotes healing, making it a valuable ingredient in skincare formulations.

When it comes to bioavailability, silymarin has relatively low absorption when taken orally, likely less than 1%.[436, 437] However, various formulations and delivery methods have been explored to enhance its bioavailability, such as using phospholipid complexes or combining it with other antioxidant compounds. A study assessed the bioavailability of various forms of silymarin and concluded that water-soluble silibinin is 50 times more bioavailable, and the micellar silibinin-phosphatidylcholine complex (phytosome) is anywhere from 10 to 40 times more bioavailable, than pure silibinin.[438, 439, 440, 441] Animal research suggests that the bioavailability of liposomal silymarin is 3.5-fold to 4.8-fold higher.[442, 443] Another form combines silibinin with proline (Silybin-L-proline cocrystal), which increases its dissolution and bioavailability (16-fold).[444] Micronized silymarin underperforms phytosome-based silymarin and nanosized silymarin produces an unimpressive 1.51–2.61-fold higher bioavailability.[445, 446] Research investigating the benefits of bioenhancers like piperine, lysergol, and fulvic acid suggests that they may enhance bioavailability, with one animal study reporting a 2.4–14.5-fold increase.[447] Some research indicates that taking silymarin with meals containing fats may also improve its absorption in the digestive tract.

A recently published meta-analysis investigated the impact of silymarin on inflammation and oxidative stress in adults.[448] This study pooled data from 16 randomized controlled trials (RCTs) to assess changes in various inflammatory and oxidative stress markers following silymarin supplementation. The findings of this meta-analysis indicated that silymarin supplementation resulted in a significant decrease in the levels of CRP and MDA. Furthermore, silymarin significantly increased the levels of IL-10, SOD, and GPx. Notably, the beneficial effects of silymarin on MDA, SOD, GPx, and CRP were more pronounced when the intervention period lasted for 12 weeks or longer. The included trials in this meta-analysis examined the effects of silymarin across a diverse spectrum of health conditions in adults, including insulin-treated diabetic patients, individuals with painful knee osteoarthritis, hemodialysis patients, patients with type 2 diabetes with and without overt nephropathy, those with homozygous (genetic) β-thalassemia major and β-thalassemia major, people with pulmonary tuberculosis, healthy subjects, people with relapsing-remitting multiple sclerosis, individuals with Hepatitis C, people with a fatty liver index between 31 and 59, individuals with endometrioma-related manifestations, and trauma patients in the Intensive Care Unit. The recency of this publication underscores its importance in providing an up-to-date assessment of the evidence. The wide range of conditions studied suggests that oxidative stress is a common thread linking these seemingly disparate diseases, and silymarin's potential impact on these markers could have broad implications.

Another systematic review and meta-analysis examined the existing evidence on the effect of silymarin consumption on inflammation and oxidative stress factors in adults.[449] This study included 15 RCTs identified through a systematic literature search up to September 2023 across several major databases. The meta-analysis revealed that silymarin consumption significantly decreased the levels of CRP, MDA, and IL-6. However, no significant effects of silymarin consumption were observed on the levels of IL-10, TAC, and GSH. Interestingly, the researchers found a significant non-linear relationship between the duration of the silymarin intervention and the changes in MDA levels. The authors concluded that silymarin could be beneficial in reducing inflammation in patients with diabetes and thalassemia by lowering MDA, a marker of oxidative stress, as well as CRP and IL-6, which are inflammatory markers. The consistency in the finding of reduced CRP and MDA across this meta-analysis and the one previously mentioned strengthens the evidence for silymarin's impact on these specific markers. The explicit mention of benefits for individuals with diabetes and thalassemia provides more focused information for these conditions, both of which are known to be associated with elevated oxidative stress.

A systematic review and dose-response meta-analysis evaluated the impact of silymarin administration on certain liver and kidney function parameters, and oxidative stress markers.[450] This analysis included 41 RCTs published up to January 2023. The pooled results indicated that silymarin supplementation led to a significant reduction in serum levels of alkaline phosphatase, alanine transaminase (ALT), creatinine, and aspartate aminotransferase (AST), along with a substantial elevation in serum glutathione. While there was a non-significant decrease in serum levels of gamma-glutamyl transferase, MDA, total bilirubin, albumin, TAC, and blood urea nitrogen in the overall analysis, subgroup analyses revealed a considerable decline in MDA and albumin serum values among participants with liver disease who received silymarin for 12 weeks or longer. These findings suggest that silymarin may improve certain liver markers, indicating potential liver protective (hepatoprotective) effects, particularly with long-term and high-dose supplementation. However, the study did not find evidence of kidney-protective effects or a significant overall impact on general oxidative stress markers. While this meta-analysis highlights silymarin's positive effects on liver function markers, the lack of a significant overall impact on general oxidative stress markers, except for MDA in a specific subgroup, suggests that silymarin's antioxidant effects might be more pronounced in the context of pre-existing liver disease.

Another systematic review and dose-response meta-analysis assessed the impacts of silymarin administration on cardiovascular risk factors..[451] This study included 33 trials with a total of 1943 participants. The meta-analysis revealed that silymarin supplementation led to a notable reduction in serum levels of fasting blood glucose, diastolic blood pressure, total cholesterol, triglycerides, fasting insulin, low-density lipoprotein, and hemoglobin A1c (HbA1c). However, no substantial differences were observed in body mass index (BMI), systolic blood pressure, CRP, body weight, and high-density lipoprotein (HDL) between the silymarin and control groups. The authors concluded that silymarin consumption reduces certain cardiometabolic syndrome (CMS) risk factors and has favorable impacts on lipid and glycemic profiles, with potential hypotensive effects. These findings extend the potential benefits of silymarin beyond liver health to include cardiovascular health, an area where oxidative stress and inflammation are also significant contributing factors.

The substantial body of research focusing on liver diseases aligns with silymarin's traditional use and suggests a strong biological basis for its potential effectiveness in this area, likely through its antioxidant and anti-inflammatory actions. However, the emerging evidence suggests it has broader implications beyond just liver function, potentially providing health benefits for a variety of conditions associated with oxidative stress.

In summary, silymarin is a complex of flavonolignans with potent antioxidant properties that help combat oxidative stress from free radicals, supporting liver health and skin integrity. Its capacity to protect against oxidative damage makes it beneficial for detoxification and overall cellular health. Given its multifaceted applications, silymarin remains a popular choice for those seeking natural support for liver function and skin health.

Bilberry Extract

Preferred Form(s): Bilberry standardized to 25%–40% anthocyanins (anthocyanosides) | *Typical Adult Dosage*: 100–750 mg/day (most commonly 160–320 mg/day) | *Companion Nutrients*: Vitamin C, vitamin E, zinc, quercetin, lutein, zeaxanthin, fish oil | *When to Take*: Anytime during the day with or without food

Bilberry (*Vaccinium myrtillus*) is a small, dark blue fruit that belongs to the berry family, closely related to blueberries. The extract from bilberry is particularly rich in anthocyanins, which are powerful plant pigments responsible for the deep color of the fruit. These anthocyanins are known for their strong antioxidant properties, allowing bilberry to effectively neutralize free radicals, including harmful ones like superoxide and hydroxyl radicals. By combatting these free radicals, bilberry helps reduce oxidative stress in the body.

One commonly reported benefit of bilberry extract is its support for eye health. Research has shown that bilberry can help protect the eyes from oxidative damage from age-related conditions such as macular degeneration and diabetic retinopathy.[452] These conditions are marked by damage to the retina, often worsened by oxidative stress, and bilberry's antioxidants help to protect and maintain healthy retinal function. Some studies also suggest that bilberry may improve night vision and overall visual acuity, making it a popular natural remedy to support eye health.

In addition to its eye benefits, bilberry extract has positive effects on cardiovascular health. It has been shown to improve circulation by strengthening the walls of blood vessels, making them less permeable and more resilient to damage.[453] This vascular support can lead to better blood flow throughout the body, which is crucial for overall cardiovascular health. Moreover, bilberry has anti-inflammatory properties that contribute to reducing inflammation in blood vessels,[454] further supporting heart health and potentially lowering the risk of cardiovascular diseases.

When it comes to form and bioavailability, bilberry extract contains bioactive anthocyanins that tend to have low bioavailability. This is largely because these compounds are water-loving rather than fat-loving, which makes it harder for them to pass through the body's lipid-rich cell membranes. In addition, anthocyanins are unstable in the changing pH conditions of the digestive tract and can be broken down by gut bacteria before they are fully absorbed.[455] It can also be consumed fresh, in jams, or as a juice. The bioavailability of anthocyanins may vary based on the form in which bilberry is consumed. For instance, extracts that are standardized for anthocyanin content are believed to offer higher bioavailability compared to the whole fruit. A pilot human study found that high amounts of anthocyanins reach the colon after consuming bilberry fruits, and the colon was a significant site of anthocyanin absorption.[456] Additionally, encapsulation of bilberry anthocyanins in whey protein improves short-term bioavailability, while encapsulation in citrus pectin increases intestinal accessibility of bilberry while traveling through the small intestine.[457] Some factors that can influence absorption include the presence of fats in the diet, which may help enhance the solubility of these compounds, and the individual's digestive health.

A systematic review of 28 randomized controlled trials (RCTs) evaluated the impact of berry consumption on oxidative stress biomarkers.[458] Of the approximately 56 oxidative stress biomarkers evaluated in the review, 32% showed statistically significant beneficial results. Among the studies analyzed, bilberry freeze-dried powder was associated with a significant decrease in oxidized low-density lipoprotein (LDL) levels, indicating a potential reduction in oxidative stress, and suggesting potential usefulness in individuals with elevated cholesterol levels. However, other studies using different berry extracts did not show significant differences. The conclusion of the review was that additional well-designed and longer-term studies are necessary to evaluate the effectiveness of bilberry on oxidative stress.

A review examined the impact of bilberry consumption or supplementation on metabolic and cardiovascular risk factors.[459] The analysis indicated that bilberry's antioxidant properties might benefit conditions like type 2 diabetes and cardiovascular diseases. Most of the studies involving supplements used a dose of 160 to 320 mg of bilberry anthocyanins. However, the authors emphasized the need for large-scale clinical trials to confirm these potential benefits.

A randomized, double-blind, placebo-controlled cross-over study assessed the effects of short-term bilberry supplementation (1.4 g/day of extract for 4 weeks) on glycemic control, lipid profile, antioxidant, and inflammatory status in Chinese individuals with type 2 diabetes.[460] The results showed no significant changes in body weight, blood pressure, lipid profile, antioxidant status, or inflammatory markers. There was a non-significant reduction in HbA1c levels. However, this study was likely limited due to its short duration. Research suggests that twelve or more weeks of bilberry supplementation is required to combat oxidative stress in many conditions. As such, the study indicates that a longer treatment period might be necessary to observe significant effects.

In a randomized clinical trial involving aged men with subjective memory impairment, participants consumed bilberry/red grape juice for nine weeks.[461] While no significant effects were observed on visual memory and psychomotor tempo (the overall speed and rhythm of mental and physical activities), there was a decrease in biomarkers related to tissue damage and inflammation. This suggests potential benefits in reducing chronic inflammation and oxidative stress, but since it combined bilberry and red grape juice—which contains bioactive polyphenols—the results cannot be fully attributed to bilberry anthocyanins.

According to a review of preclinical and clinical studies, anthocyanins from bilberry may have anti-inflammatory effects.[462] Data from the cell and animal models demonstrated that bilberry reduces inflammation by lowering TNF-α, IL-1β, iNOS, and COX, and altering NF-κB and JAK/Stat signaling pathways. A clinical study showed that supplementation with 320 mg of pure anthocyanins (equal to about 100 g of fresh bilberries) daily for four weeks significantly reduced inflammation in subjects with metabolic syndrome,[463] as evidenced by decreased levels of inflammatory biomarkers and improved lipid profiles. The researchers hypothesized that consumption of bilberry may beneficially alter the gut microbiome, producing an anti-inflammatory effect.

In summary, bilberry extract is a potent source of anthocyanins that provides strong antioxidant activity against free radicals, making it beneficial for protecting eye health and supporting cardiovascular function. Its positive effects on reducing oxidative stress position bilberry as a valuable addition to a healthy lifestyle, particularly for visual acuity and cardiovascular health. With various forms available, bilberry can easily be incorporated into daily routines to harness its health benefits.

Lycopene

Preferred Form(s): Lycopene softgel | *Typical Adult Dosage*: 15–60 mg/day (most commonly 20 mg/day) | *Companion Nutrients*: Vitamin C, vitamin E, zinc, selenium, quercetin, lutein, fish oil | *When to Take*: Anytime during the day with food

Lycopene is a vibrant red carotenoid pigment that is primarily found in tomatoes, as well as in other red and pink fruits and vegetables such as watermelon, pink grapefruit, and red peppers. This powerful antioxidant is well-known for its ability to neutralize free radicals, particularly singlet oxygen and hydroxyl radicals, which are highly reactive molecules that can cause significant cellular damage and contribute to oxidative stress. By scavenging these harmful radicals, lycopene plays a crucial role in protecting cells from oxidative damage, thereby supporting overall health and well-being.

Research has established a compelling link between lycopene intake and a reduced risk of several chronic diseases, particularly prostate cancer and cardiovascular diseases.[464,465] In terms of prostate health, multiple studies have suggested that higher dietary intake of lycopene is associated with a lower risk of prostate cancer, possibly due to its ability to inhibit cancer cell growth and reduce inflammation in the prostate gland. Similarly, lycopene's anti-inflammatory properties can help improve cardiovascular health by reducing arterial stiffness, lowering blood pressure, and decreasing the oxidation of LDL cholesterol, a key factor in heart disease.

When it comes to bioavailability, the absorption of lycopene can vary based on how it is consumed. Interestingly, the bioavailability of lycopene is generally higher when tomatoes are processed or cooked, such as in tomato sauce or ketchup, rather than when they are eaten raw. Cooking tomatoes breaks down the cell walls and makes the lycopene more accessible to the body. Additionally, lycopene is fat-soluble, meaning that its absorption is enhanced when consumed with dietary fats, such as olive oil or avocados. Most of the lycopene found in foods is trans-lycopene, but cis-lycopene seems to be better absorbed due to its shorter length and better solubility.[466] The human body rapidly and significantly alters the trans-lycopene form to cis-lycopene after absorption.[467] This is why incorporating healthy fats into meals that include tomatoes can maximize the beneficial effects of lycopene.

Lycopene is available as a dietary supplement in various forms, including softgels and capsules, often derived from concentrated tomato extract. When used in supplement form, about 17%–29% of pure lycopene is absorbed by humans.[468] However, its absorption can be significantly affected by genetic individuality, or single nucleotide polymorphisms (SNPs), such as BCO1 SNPs (rs6564851 is associated with lower serum lycopene levels; rs12934922 also influences serum lycopene concentrations), and SCARB1 SNPs (variants are linked to lower serum lycopene). MTTO SNPs, SOD2 SNPs, and elongation of very long-chain fatty acid protein 2 also significantly affect lycopene absorption. Some lycopene supplements come in beadlet form that is microencapsulated with vitamins E or C to improve stability. But whether this aids absorption and bioavailability is not fully known.

The role of lycopene in promoting cardiovascular health has been a subject of considerable research, with several studies evaluating its impact on key risk factors and related conditions. A meta-analysis encompassing six intervention trials investigated the effect of lycopene supplementation on blood pressure.[469] The pooled analysis demonstrated a statistically significant reduction in systolic blood pressure (SBP) among participants who received lycopene compared to the control groups. Notably, this effect was more pronounced with higher dosages of lycopene supplementation, exceeding 12 mg per day. Subgroup analysis revealed that individuals with a baseline SBP greater than 120 mmHg and participants of Asian descent experienced a more significant decrease in SBP. This suggests that lycopene's impact on blood pressure may be influenced by both the dosage administered and the characteristics of the population being treated. However, the same meta-analysis found that lycopene intervention did not have a statistically significant effect on diastolic blood pressure (DBP). This indicates a potentially selective action of lycopene on SBP regulation. Further supporting these findings, a separate review highlighted lycopene's ability to reduce blood pressure through mechanisms such as inhibiting the angiotensin-converting enzyme (ACE) and enhancing the bioavailability of nitric oxide, a molecule crucial for vasodilation.[470] The evidence suggests that lycopene might serve as a valuable adjunct in managing hypertension, particularly systolic hypertension, although further research is needed to fully elucidate the optimal dosage and target populations for this benefit.

Beyond blood pressure, lycopene has been associated with protection against coronary artery disease (CAD) and the underlying process of atherosclerosis. Multiple reviews suggest that lycopene can positively influence lipid profiles, a key factor in atherosclerosis development.[471,472,473,474,475] Specifically, it has been shown to potentially lower levels of low-density lipoproteins (LDL) and improve levels of high-density lipoproteins (HDL), thereby contributing to the minimization of atherosclerotic plaque formation— the buildup of fatty, sticky material along the walls of arteries, which narrows them and makes it harder for blood to flow. The mechanisms through which lycopene exerts these protective effects are multifaceted. It can suppress the proliferation of vascular smooth muscle cells and the formation of foam cells, both critical events in atherogenesis—a chronic inflammatory disorder characterized by endothelial dysfunction and the accumulation of several cellular components in the inner layer of the blood vessels. Furthermore, lycopene's potent antioxidant properties enable it to inhibit the oxidation of LDL, a process that significantly contributes to the development of atherosclerotic lesions. Additionally, lycopene has been shown to reduce inflammation, another pivotal factor in the progression of CAD. A comprehensive review further emphasized lycopene's broad impact on vascular health, encompassing antiatherosclerotic, antioxidant, anti-inflammatory, antihypertensive, antiplatelet, and anti-apoptotic effects, as well as its ability to protect the endothelium, improve metabolic profiles, and reduce arterial stiffness. The convergence of these findings across various reviews underscores the significant potential of lycopene in both the prevention and management of cardiovascular diseases.

The influence of lycopene on lipid profiles and endothelial function further supports its role in cardiovascular health. Research indicates that lycopene can contribute to the normalization of lipid metabolism by reducing total serum cholesterol, triglycerides, and LDL-cholesterol levels. Moreover, it has been consistently shown to improve endothelial function, which is essential for maintaining the health and integrity of blood vessels. Observational studies have also provided valuable insights. For instance, higher concentrations of carotenoids in the serum, including lycopene, were associated with a lower risk of both all-cause and cardiovascular mortality among adults with hypertension.[476] Similarly, higher serum lycopene concentrations were found to be independently linked to a decreased risk of all-cause and cardiovascular disease mortality in individuals with chronic kidney disease.[477] These findings suggest that maintaining adequate levels of lycopene may be particularly beneficial for individuals at higher risk of cardiovascular events due to pre-existing conditions. The combined evidence from meta-analyses, reviews, and observational studies points towards a significant role for lycopene in mitigating cardiovascular risk through its positive effects on blood pressure, atherosclerosis, lipid profiles, and endothelial function.

The potential role of lycopene in cancer prevention and management has been researched extensively. A comprehensive meta-analysis of prospective cohort studies investigated the association between dietary intake and blood levels of lycopene and the risk of developing both overall and specific types of cancer.[478] The findings revealed that higher intakes and blood concentrations of lycopene were associated with a statistically significant reduction in the risk of overall cancer. Specifically, individuals with high lycopene intake experienced a 5% lower risk, while those with high blood levels of lycopene showed an 11% lower risk of overall cancer. Furthermore, the analysis indicated that for every 10 mcg/dL increase in blood lycopene levels, there was a corresponding 5% decrease in the risk of overall cancer. To put this in perspective, on average, it would require consuming between 200 and 250 mg of lycopene over about a week to increase levels by 10 mcg/dL. This meta-analytic evidence strongly suggests a protective association between higher lycopene exposure and a reduced overall likelihood of developing cancer, highlighting its potential as a dietary component for cancer prevention strategies.

In addition to the overall cancer risk, the meta-analysis also examined the relationship between lycopene and specific cancer types. Higher intakes and blood levels of lycopene were found to be negatively associated with the risk of lung, breast, and prostate cancers. Notably, a significant inverse association was observed between lycopene intake and the risk of lung cancer. Similarly, higher blood levels of lycopene were significantly associated with a reduced risk of both breast and prostate cancer. This is interesting since lycopene is known to preferentially concentrate in the liver and prostate, with lower accumulation in the lungs, breasts, testes, ovaries, kidneys, and skin.[479] These findings are further supported by a systematic review that specifically focused on the anti-cancer activities of lycopene in both human and animal studies, which confirmed its potential role, particularly in the context of prostate cancer.[480] Moreover, research suggests that lycopene may play a role in reducing the incidence of gastric cancer through its antioxidant mechanisms and its ability to modulate cell proliferation and apoptosis, key processes in cancer development.[481] While the evidence for a broad protective effect of lycopene against cancer is growing, these specific associations with lung, breast, prostate, and gastric cancers warrant further in-depth investigation to fully understand the underlying mechanisms and the potential for targeted interventions. The consistency of findings across different types of studies, including large-scale cohort studies and *in vivo* research, for certain cancers like prostate cancer is particularly noteworthy and suggests a promising area for future research.

The influence of lycopene on metabolic disorders, particularly diabetes and metabolic syndrome, has also been a focus of scientific inquiry. A meta-analysis that examined the effect of lycopene intake on fasting blood glucose (FBG) levels suggested that lycopene consumption may have an FBG-lowering effect, especially in individuals with type 2 diabetes.[482] Dosing ranged from 6 to 50 mg/day with intake periods spanning from two weeks to six months. This finding indicates a potential role for lycopene in glucose metabolism. Furthermore, a separate review highlighted that lycopene shows promise in both preventing and treating type 2 diabetes.[483] This potential benefit is attributed to lycopene's ability to lower blood glucose levels, improve the body's sensitivity to insulin, and mitigate the oxidative stress that is often associated with diabetes. Mechanistically, it has been proposed that lycopene may enhance insulin signaling pathways within the body, specifically through the activation of the PI3K/AKT pathway, which plays a crucial role in glucose and lipid metabolism.[484] While the existing evidence suggests that lycopene may have beneficial effects on glucose regulation, particularly in the context of type 2 diabetes, further research is necessary to definitively confirm these findings, determine the optimal dosages for therapeutic effects, and understand the long-term implications of lycopene supplementation in managing diabetes.

Beyond diabetes, lycopene's antioxidant and anti-inflammatory properties have positioned it as a promising agent in the prevention and treatment of metabolic syndrome and obesity. Observational studies have indicated that higher levels of lycopene in the blood are associated with a lower risk of developing metabolic syndrome.[485,486] Animal studies from these reviews provide further support, suggesting that lycopene can improve lipid profiles, lower glucose and insulin levels, and reduce the oxidative stress and inflammation that are commonly observed in obesity. These findings suggest that lycopene may exert a positive influence on multiple interconnected components of metabolic syndrome and obesity. This is likely due to its ability to counteract oxidative damage and reduce inflammation, both of which are key underlying factors in the development of these conditions. While human studies in this area are still evolving, the evidence from animal models, coupled with some human observational data, suggests a potential role for lycopene in managing metabolic syndrome and obesity.

The possibility of lycopene mitigating the effects of neurodegenerative diseases has garnered attention, primarily due to its potent antioxidant properties. Conditions such as Alzheimer's disease (AD) and Parkinson's disease are characterized by significant oxidative stress within the brain, leading to neuronal damage and dysfunction. Given lycopene's well-established ability to scavenge free radicals and reduce oxidative damage, it has been hypothesized that it could offer neuroprotective benefits. A review of the literature highlighted numerous reports suggesting the potential application of lycopene in the treatment of AD and other neurodegenerative conditions.[487] This interest stems from lycopene's demonstrated antioxidant and cell-protective activities, which could help to counteract the damage caused by amyloid-beta plaques and potentially enhance cognitive function. While preclinical studies and *in vitro* experiments have provided encouraging results regarding lycopene's neuroprotective potential, more human clinical trials are needed to definitively establish its efficacy in treating or preventing neurodegenerative diseases.

Beyond the major categories of cardiovascular health, cancer, and metabolic and neurodegenerative disorders, research has also explored lycopene's effects on other conditions linked to oxidative stress. A meta-analysis involving seven studies specifically examined the antioxidant effect of topically applied lycopene as an adjunct treatment for periodontal disease (PD).[488] The combined data from these studies indicated that lycopene treatment, when used alongside conventional periodontal therapy, resulted in a beneficial antioxidant effect on PD. Statistically significant improvements were observed in several clinical parameters, including a reduction in plaque index in the short, medium, and long term. Additionally, lycopene treatment led to noteworthy improvements in bleeding on probing and gingival index during the short and medium follow-up periods. Furthermore, research suggested that lycopene could increase the levels of uric acid, an antioxidant present in saliva. This increase in salivary antioxidant capacity may contribute to slowing down the destruction of periodontal tissues that is a characteristic of PD. The evidence from this meta-analysis suggests that lycopene may serve as a valuable complementary therapy in managing periodontal disease, likely through its antioxidant properties that help to improve gum health and reduce inflammation.

Another area of investigation involves the potential of lycopene to protect against skin damage induced by ultraviolet (UV) radiation. One study concluded that a lycopene-rich tomato nutrient complex (TNC) supplement provided significant protection against the skin redness and the upregulation of pro-inflammatory cytokines in healthy volunteers exposed to UVB radiation.[489] Participants consumed the TNC containing 15 mg lycopene, 5.8 mg phytoene and phytofluene, 5.6 mg tocopherols, 4 mg carnosic acid, and 0.8 mg beta-carotene, or a placebo for twelve weeks. This finding suggests that lycopene, as part of a broader nutrient complex, may offer a degree of protection against the oxidative stress and inflammation that results from UV exposure in the skin. This protective effect could have implications for mitigating the harmful effects of sunlight on the skin and potentially reducing the risk of long-term damage associated with UV radiation.

In summary, lycopene is a powerful antioxidant found primarily in tomatoes that effectively protects against oxidative stress by neutralizing harmful free radicals. Its association with reduced risks of prostate cancer and cardiovascular diseases highlights its importance in a health-conscious diet. By understanding the best ways to consume and absorb lycopene, individuals can take advantage of its many health benefits, making it an important addition to a balanced diet rich in colorful fruits and vegetables.

In a nutshell, there are multiple powerful antioxidant supplement options to help combat oxidative stress. Based on the majority of the research, supplementation should continue for at least twelve weeks for measurable results. Many of the antioxidants in this chapter can be taken together to improve results as they work synergistically, positively affect bioavailability, or recycle one another. It is important to use high-quality supplements that have been independently tested for their active ingredients.

CHAPTER SEVEN

Essential Oils as Antioxidants

When considering essential oils and their antioxidant protection, clinical data is limited, which necessitates a reliance on lab tests that measure specific biological activity evaluating their antioxidant capacity, along with animal studies that demonstrate their protective properties. These laboratory assays often measure the ability of essential oils to neutralize free radicals and scavenge reactive oxygen species, providing insight into their potential health benefits. Additionally, animal studies can help elucidate the mechanisms by which these oils exert their protective effects, paving the way for future research aimed at validating their use in human health.

Despite the lack of clinical evidence, some essential oils exhibit potent antioxidant activity. Their effectiveness compared to other antioxidants arises from a unique combination of chemical structure, bioavailability, and mechanisms of action. Essential oils have a distinct advantage due to their lipophilic (fat-loving) constituents, such as terpenes, phenols, and aldehydes. Cell membranes are lipid-based, so lipophilic molecules in essential oils can penetrate and integrate into these membranes more effectively than hydrophilic (water-loving) antioxidants. Additionally, the mitochondria, a major source of oxidative stress, are surrounded by lipid membranes. Lipophilic constituents can enter a cell and reach these organelles more easily and may exert protective effects directly at the site of reactive oxygen species (ROS) generation. This means they can go straight to where damaging molecules are being produced and help neutralize them before they spread and cause further harm.

Essential oils are not single compounds but complex mixtures of dozens to hundreds of volatile organic molecules. This complexity often leads to synergistic interactions, where the combined antioxidant effect is greater than the sum of individual components. For example, monoterpenes (like limonene), phenols (like eugenol in clove), and sesquiterpenes (like beta-caryophyllene) may act on multiple antioxidant pathways simultaneously.

Unlike traditional antioxidants that primarily act as a direct ROS scavenger, essential oils provide broad-spectrum antioxidant protection through multiple, targets, pathways, and mechanisms. They can upregulate endogenous antioxidant defenses like GPX, SOD, and CAT, modulate cellular signaling pathways (e.g., Nrf2) that govern oxidative stress responses, and offer both preventive and interventional effects—protecting cells from oxidation and also assisting in the recovery of damaged systems.

The volatile nature and molecular size of essential oils also allows for diffusion across barriers—like the blood-brain barrier—allowing access to privileged tissues. Diffusion is the natural movement of molecules from an area where there are a lot of them to an area where there are fewer of them—until they spread out evenly. Think of diffusing an essential oil in one corner of a room. At first, the scent is strongest where your diffuser is located. But soon, the scent spreads throughout the room without you doing anything. That spreading happens because of diffusion—the aromatic molecules move from where they're crowded (high concentration) to where there are fewer of them (low concentration).

Volatility means these compounds easily evaporate and become airborne, allowing for rapid diffusion into biological systems via inhalation, topical application, and even ingestion. Their small molecular size and lipophilicity enable them to translocate through biological membranes, including epithelial layers, vascular endothelium, and cellular lipid bilayers. So, essential oils can affect tissues that are historically difficult to get to such as the brain and lower respiratory tract. The blood-brain barrier (BBB) is a selective, tightly regulated filter that protects the central nervous system. Many water-soluble antioxidants (like vitamin C) have limited access unless specialized transport mechanisms are available. Contrarily, essential oils readily diffuse across the BBB, entering the brain within minutes of inhalation. Once there, they can direct antioxidant and neuroprotective effects in brain tissue, where oxidative stress plays a role in neurodegenerative diseases (e.g., Alzheimer's, Parkinson's, MS). Inhalation of essential oils is also an ideal way to target oxidative stress in the lungs, which is seen in conditions like asthma and Chronic Obstructive Pulmonary Disease (COPD). You also have some systemic delivery through the lungs via the highly vascularized olfactory and respiratory epithelium, offering a non-invasive delivery route. Inhaled essential oils can exert antioxidant effects before first-pass metabolism, unlike oral compounds that must survive digestion and liver metabolism. The olfactory system also provides a direct neural pathway to the limbic system, where essential oils impact mood, stress, and neuroinflammation.

Due to their volatile and lipophilic nature, essential oil compounds disperse quickly through tissues, are absorbed into the bloodstream and distributed to lipid-rich organs (brain, liver, adrenal glands), and have a fast onset of action. When inhaled, volatile constituents can enter the bloodstream via the lungs and also interact directly with the brain through the olfactory bulb, influencing mood and neuroendocrinological pathways. When applied topically, their lipophilicity enables penetration through the skin and absorption into local capillaries, allowing both localized and systemic effects. When ingested, essential oils are absorbed in the gastrointestinal tract—often via passive diffusion—and, depending on the formulation, may enter either the bloodstream or lymphatic system, facilitating delivery to internal tissues. Each route offers distinct advantages for therapeutic targeting, making essential oils uniquely versatile in supporting physiological function and cellular health.

The following is a list of common essential oils that have exhibited antioxidant activity in laboratory assays and how that might be applicable to free radicals that humans are exposed to in the real world. They will be divided based on their activity in specific biological systems and tissues based on the test performed to determine their antioxidant capacity.

DPPH (2,2-diphenyl-1-picrylhydrazyl) Assay

What it measures: Ability of a substance to donate a hydrogen atom to neutralize a stable DPPH radical.

Real-world free radical: Environmental pollutants or cigarette smoke introducing free radicals that damage skin cells. These are also produced during normal metabolism, inflammatory processes, and when exposed to UV radiation.

Essential oils: Ajowan, allspice, basil CT linalool, basil (tropical), bay laurel, bergamot, mint, caraway, cardamom, cassia, cedarwood (Atlas), celery seed, cinnamon bark, cinnamon leaf, clary sage, clove, coriander, cumin, cypress, eucalyptus, fennel, frankincense, geranium, German chamomile (multiple CTs), ginger, grapefruit, helichrysum, hemp, juniper, lavandin, lavender, lemon, lemongrass, lemon eucalyptus, lemon myrtle, lemon verbena, lime, marjoram, may chang (litsea), mountain (winter) savory, neroli, nutmeg, orange (sweet/wild), oregano, palmarosa, patchouli, peppermint, petitgrain, pink pepper, ravintsara, Roman chamomile, rose, rosemary, sage, Spanish sage, spearmint, tangerine, tea tree, thyme, turmeric, ylang ylang, and yuzu.

ABTS/TEAC (2,2'-azino-bis (3-ethylbenzothiazoline-6-sulfonic acid)/Trolox Equivalent Antioxidant Capacity)

What it measures: Electron donation to neutralize ABTS•+ radicals.

Real-world free radical: Oxidative stress from UV exposure leading to skin aging, air pollution, cigarette smoke, chronic inflammation, and mitochondrial metabolism.

Essential oils: Allspice, basil CT linalool, basil (tropical), cardamom, cassia, cedarwood (Atlas), cinnamon bark, cinnamon leaf, clary sage, clove, coriander, cypress, eucalyptus, fennel, frankincense, geranium, ginger, grapefruit, helichrysum, hemp, lavender, lemon, lemongrass, lime, mandarin, marjoram, may chang (litsea), orange (sweet/wild), oregano, palmarosa, peppermint, petitgrain, pink pepper, ravintsara, rosemary, rose, rosewood, sage, spearmint, tangerine, tea tree, thyme, turmeric, ylang ylang, and yuzu.

FRAP (Ferric Reducing Antioxidant Power)

What it measures: Ability to reduce ferric (Fe^{3+}) to ferrous (Fe^{2+}) ions, indicating reducing power.

Real-world free radical: Iron overload or inflammation generating hydroxyl radicals or produced during metabolic stress and immune system activation.

Essential oils: Ajowan, basil CT linalool, bay laurel, cardamom, cedarwood (Atlas), cinnamon bark, caraway, clove, coriander, cumin, cypress, eucalyptus, geranium, ginger, grapefruit, hemp, lavender, lemongrass, lemon myrtle, mandarin, marjoram, may chang (litsea), palmarosa, peppermint, petitgrain, orange (sweet/wild), oregano, rosemary, sage, Spanish sage, tangerine, thyme, and ylang ylang.

H2O2 (Hydrogen Peroxide Scavenging) Assay

What it measures: The ability of a compound or extract to scavenge hydrogen peroxide or prevent its accumulation.

Real-world free radical: ROS formed by mitochondrial activity and various oxidative enzymes, as well as inflammatory and immune responses.

Essential oils: Black pepper, basil (tropical), cassia, citronella, clove, eucalyptus, ginger, lavender, lemon, lemon verbena, lemongrass, marjoram, myrrh, palmarosa, rosemary, sage, tea tree, and thyme.

ORAC (Oxygen Radical Absorbance Capacity)

What it measures: Capacity to scavenge peroxyl radicals using fluorescein as a probe.

Real-world free radical: Peroxyl radicals formed during normal metabolism or inflammation.

Essential oils: Basil, citronella, clove, eucalyptus, juniper, oregano, pine, rosemary, sage, and thyme.

HORAC (Hydroxyl Radical Averting Capacity)

What it measures: Metal-chelating antioxidant capacity to prevent hydroxyl radical formation via Fenton reaction.

Real-world free radical: Hydroxyl radicals generated from pollution or radiation causing DNA strand breaks.

Essential oils: Anise, bay laurel, cinnamon bark, clove, eucalyptus, lemongrass, may chang (litsea), pink pepper, nutmeg, spearmint, tea tree, and ylang ylang.

Beta-carotene Bleaching Assay

What it measures: Lipid peroxidation damaging cell membranes (e.g., in the liver or brain).

Real-world free radical: Ability of antioxidants to prevent oxidation of β-carotene by linoleic acid-derived free radicals.

Essential oils: Bay laurel, black pepper, caraway, cumin, eucalyptus, fennel, lemongrass, orange (sweet/wild), oregano, nutmeg, pink pepper, ravintsara, rosemary, and sage.

TBARS (Thiobarbituric Acid Reactive Substances)

What it measures: Levels of MDA as a marker of oxidative damage to lipids.

Real-world free radical: Malondialdehyde (MDA) production from lipid peroxidation during high-fat diet or alcohol abuse.

Essential oils: Clove, fennel, nutmeg, oregano, peppermint, sage, and thyme.

NO (Nitric Oxide) Scavenging Assay

What it measures: Ability of compounds to scavenge nitric oxide or reduce nitrite accumulation, which becomes reactive under oxidative conditions.

Real-world free radical: Elevated NO during inflammation contributes to tissue damage in conditions like arthritis, sepsis, or neurodegeneration.

Essential oils: Cinnamon bark, clove, copaiba, coriander, helichrysum, oregano, peppermint, rosemary, tea tree, thyme, yarrow, and ylang ylang.

XO (Xanthine Oxidase) Inhibition Assay

What it measures: Inhibition of the xanthine oxidase enzyme, which catalyzes the conversion of xanthine to uric acid, generating ROS.

Real-world free radical: Gout (XO overactivity leads to uric acid buildup), cardiovascular and renal oxidative damage.

Essential oils: Cinnamon bark, clove, ginger, oregano, peppermint, pink pepper, and rosemary.

CAA (Cellular Antioxidant Activity)

What it measures: Intracellular antioxidant defense in living cells against free radical-induced oxidative stress.

Real-world free radical: Cellular ROS generation during inflammation or infection.

Essential oils: Clove, frankincense, orange (sweet/wild), lavender, oregano, rosemary, and thyme.

SOD (Superoxide Dismutase) Activity

What it measures: Enzymatic activity to break down (dismutate) superoxide into safer molecules—oxygen and hydrogen peroxide.

Real-world free radical: Superoxide radicals formed in mitochondria during normal respiration.

Essential oils: Bergamot, cinnamon bark, clove, fennel, frankincense, ginger, lavender, lemongrass, myrrh, myrtle, oregano, peppermint, rosemary, thyme, and tea tree.

CAT (Catalase Activity)

What it measures: Catalase enzyme efficiency in converting hydrogen peroxide to water and oxygen.

Real-world free radical: Hydrogen peroxide buildup from oxidative stress.

Essential oils: Bergamot, black pepper, fennel, ginger, lavender, myrrh, myrtle, rosemary, and thyme.

TAC (Total Antioxidant Capacity)

What it measures: Overall ability of biological fluids (like plasma or serum) to neutralize free radicals.

Real-world free radical: General oxidative stress from poor diet, pollution, or disease.

Essential oils: Ajowan, bergamot, cassia, oregano, rosemary, and thyme.

GPx (Glutathione Peroxidase) Activity Assay

What it measures: The enzymatic activity of glutathione peroxidase, which catalyzes the reduction of peroxides using reduced glutathione (GSH).

Real-world free radical: Crucial for detoxifying endogenous peroxides in organs like the liver, kidneys, and brain.

Essential oils: Fennel, lavender, myrrh, myrtle, patchouli, peppermint, rosemary, and thyme.

GSH (Glutathione Status) Assay

What it measures: The concentration or redox status of glutathione: GSH (reduced form)—the active antioxidant, GSSG (oxidized form)—indicates oxidative stress, and GSH:GSSG ratio—a biomarker of cellular oxidative balance.

Real-world free radical: One of the most critical intracellular antioxidants that protects against chronic diseases (e.g., cancer, CVD, HIV), toxic effects of drugs and environmental chemicals, and even slows the aging process.

Essential oils: Anise, bergamot, black pepper, citrus oils high in limonene, clove, ginger, German chamomile, eucalyptus (*E. globulus*), frankincense, lavandin, lavender, lemongrass, rosemary, and spearmint.

Together, these laboratory tests provide a comprehensive framework for understanding how antioxidants work at the cellular level and how their effects are measured in controlled settings. While assays help scientists identify antioxidant potential and mechanisms of action, the real question is how these effects translate into living systems and meaningful health outcomes. The following examples move beyond test results to illustrate how specific essential oils—validated through these antioxidant assays—have demonstrated protective, anti-inflammatory, and antioxidant effects in biological models, offering insight into their potential real-world relevance.

Prostaglandins play an important role in primary dysmenorrhea (PD; painful menstruation including cramping in the lower abdomen at the onset of menstruation without an identifiable cause). Research suggests that women who experience severe menstrual pain have higher levels of prostaglandin PGF2alpha and PGE2. Administration of bergamot essential oil—via gastric perfusion—reduced symptoms of PD (pain and uterine tissue alterations) in rats by reducing PGF2alpha and PGE2 levels, decreasing the inflammatory response (inhibiting the release of inducible nitric oxide), and diminishing oxidative stress in uterine tissues (reducing MDA and increasing antioxidant activities—T-AOC, SOD, CAT, and GSH).[490] The essential oil was more effective than bergamot juice and ethanol extract. Another study concluded that bergamot essential oil checks oxidative stress in brain neuronal cells and potentially prevents neurodegenerative conditions.[491] A third study showed that bergamot oil protects against oxidative stress induced by strenuous physical work, excessive exercise, or continuous long-term study or physical activity, protecting muscles against injury caused by free radicals.[492] Bergamot appears to have great promise in protecting against damage caused by oxidative stress and warrants further research in human clinical trials.

Alzheimer's disease (AD) is the most prevalent type of dementia, and the number of people living with it is expected to triple by 2050. The progressive memory loss and behavioral changes over years is a painful process for those with AD and their families and caregivers. A randomized controlled clinical trial investigated the effects of aromatherapy—sleep quality, neuropsychiatric behavioral disorders, quality of life, and biochemical measures—for people with AD.[493] Six drops of a blend of lavender, sweet orange, and bergamot essential oils (2 drops of each) were added to 200 mL of water in a diffuser, which was operated every night at 9:00 p.m. for one hour over a 12-week period. The control group inhaled from a diffuser with water only. Inhaling the essential oil blend improved sleep quality, alleviated psychobehavioral symptoms, and improved overall quality of life. Biochemical evaluation also showed a decrease in oxidative stress (by increasing SOD levels) and inflammation (decreased TNF-α and IL-6 levels). The biochemical results are noteworthy because oxidative stress and inflammation are linked to further deterioration of AD.

Animal research suggests that caraway essential oil protects the kidneys from oxidative stress caused by sepsis (a life-threatening complication of an infection that occurs when chemicals released into the bloodstream to fight infections trigger a systemic inflammatory response).[494] This protective effect was observed by measuring certain substances—TBARS, GSH, urea, and creatinine—in the kidney and blood, indicating that the oil can help maintain normal kidney function.

Pulmonary fibrosis is a lung disease in which the lungs become scarred over time, leading to shortness of breath and dry cough. Researchers evaluated the effects of citronella (*Cymbopogon winterianus*) essential oil in a rat model of pulmonary fibrosis. Oral administration of citronella oil significantly reduced inflammatory markers in bronchoalveolar lavage fluid, reduced signs of oxidative stress (malondialdehyde), and increased antioxidant enzyme (superoxide dismutase) activity.[495]

When rats were given clove essential oil orally, it remarkably counteracted the harmful effects of oxidative stress. This was observed in two primary outcomes: by improving tissue health and by balancing blood chemistry. Regarding tissue health (histopathological changes), the oil improved conditions in several organs. In the liver, it reduced the swelling of central veins, inflammation, and the presence of abnormal cells with two nuclei (often seen in cancer). In the brain, it lessened congestion and bleeding. In the kidneys, it decreased congestion. As for blood chemistry (biological changes), the treatment also brought various blood parameters back to normal levels, including blood urea, liver enzymes, and different fat levels like total cholesterol, LDL, VLDL, HDL, and triglycerides. Additionally, it normalized total protein and albumin levels.[496] The research suggests that the antioxidant properties of clove protect the liver, brain, and kidneys against toxicity.

Pulmonary arterial hypertension is high blood pressure that affects the arteries in the lungs and the right side of the heart. It occurs when these blood vessels become narrowed, blocked, or damaged. It is characterized by shortness of breath, chest pressure, and dizziness that worsens over time. Chronically elevated pulmonary arterial pressure can cause right ventricle hypertrophy (abnormal enlargement and thickening of the walls of the right ventricle) and increase pulmonary vascular resistance (resistance against blood flow from the pulmonary artery to the left atrium, which makes the right ventricle work harder to move blood through pulmonary circulation). Oxidative stress in cardiovascular tissue is an early contributing factor to pulmonary hypertension. Oral administration of copaiba essential oil (400 mg/kg BW) daily for one week reduced pulmonary arterial hypertension in rats.[497] Rats were divided into six groups: (1) control that received no treatment and absent of hypertension; (2) copaiba essential oil; (3) copaiba nanocapsules; (4) medium chain triglycerides (MCT) to trigger oxidative stress, pulmonary vascular resistance, and right ventricle hypertrophy; (5) copaiba oil + MCT; and (6) copaiba nanocapsules + MCT. Echocardiograms were performed to assess cardiovascular effects. The researchers found that both copaiba oil and copaiba nanocapsules significantly reduced right ventricle hypertrophy and oxidative stress, but only copaiba oil reduced pulmonary vascular resistance. Taken together, these findings suggest that by reducing oxidative stress, copaiba essential oil may help protect the heart and lung blood vessels from the structural and functional strain that drives pulmonary hypertension—at least in experimental models.

Animal research suggests that the inhalation of coriander oil may improve spatial memory (ability to recall information about one's environment) and reduce oxidative stress of the hippocampus in rats with cognitive impairment.[498] The researchers noted changes in the expression of Nrf2 and p21, which are considered to be important players in response to oxidative stress.

The incidence of metabolic syndrome (MetS), a cluster of connected serious metabolic disorders, is increasing and represents a major health problem associated with elevated oxidative stress and inflammation. A randomized, triple blind, placebo-controlled clinical trial found that oral administration of 75 mg of cumin essential oil, three times daily, for eight weeks, reduced markers of MetS in people aged 18 to 60 years.[499] The oil improved antioxidant activity and simultaneously decreased free radicals, therefore reducing oxidative stress.

Oral administration of fennel or rosemary oils (or their respective nanoforms) in dyslipidemic rats improved biochemical parameters (reduced cardiovascular disease risk, decreased inflammation and oxidative stress, preserved liver function, and lowered blood sugar levels).[500] This suggests that these oils, especially in nano-form, can protect against cell damage caused by oxidative stress.

The widespread use of pesticides is a concern for humans and the environment. Researchers found that fennel essential oil protects against pesticide (triflumuron) toxicity in laboratory research.[501] The oil increased cell viability and antioxidant (SOD and CAT) activity, reduced ROS generation, and significantly decreased DNA damage and malondialdehyde levels.

A study looked at geranium essential oil and its effects on diabetic rats.[502] The scientists found that this oil has antioxidant properties, which means it can help protect cells from damage associated with diabetes. In diabetic rats, these harmful molecules are frequently elevated, leading to oxidative stress. The essential oil seemed to reduce oxidative stress by boosting the body's natural defense systems against these molecules in the liver, kidneys, and blood. By acting as an antioxidant, geranium essential oil may indirectly help protect against DNA damage in diabetic conditions.

Lavender essential oil has been investigated in multiple models of oxidative stress damage. Lavender essential oil improved memory in a mouse model of

Alzheimer's disease by decreasing oxidative stress in neurons (increased GPx, CAT, and SOD).[503] Oral administration of lavender essential oil (100 mg/kg) reduced oxidative stress associated with psychological stress and reduced stress-induced behavioral and biochemical changes more effectively than ibuprofen in rats.[504] A significant decrease in plasma corticosterone and lipid peroxidation levels and an increase in antioxidant enzyme activities was observed after lavender administration. Lavender essential oil reduced cardiac injury after myocardial infarction (heart attack) in rats.[505] Intraperitoneal administration of the oil immediately after ischemia (inadequate blood supply to an organ) reduced inflammation and oxidative stress to protect the heart against damage. An experimental model of sepsis-induced acute lung injury found that ingestion of lavender essential oil protected rats against sepsis-induced acute lung injury.[506] Specifically, the oil reduced inflammation and oxidative stress (which triggers cell death) and reduced overexpression of genes linked to lung injury. Taikong blue lavender is a cultivar of lavender used to produce an essential oil in Xinjiang Province, China. Taikong blue lavender (41.0% linalyl acetate, 29.5% linalool) reduced oxidative stress indicators (NO, ROS, MDA, and iNOS at the mRNA and protein levels) and preserved antioxidant (SOD, CAT) activities in inflamed cells.[507] Lavender has multi-organ protective properties that are highly valuable.

Animal research suggests that lemon essential oil prevents damage to the hippocampus (the part of the brain involved in the storage of long-term memory, knowledge, experiences, and the related emotions) caused by oxidative stress associated with neurodegenerative diseases.[508] Specifically, the oil significantly reduced lipid (fat) peroxidation while simultaneously increasing GSH levels and the SOD, catalase, and GPx activities in the brain. Since the brain is approximately 60% fat by dry weight, making it the fattiest organ in the human body, and this fat is not storage fat but crucial for structural components of brain cells and nerve fibers, the reduction of fat oxidation in the brain preserves brain function and communication.

Lemongrass (*Cymbopogon flexuosus*) and citral both prevented oxidative stress in the liver caused by acetaminophen at an oral dose of 400 mg/kg body weight.[509] The study determined that lemongrass and citral modify drug metabolizing enzymes but do not affect the detoxification of acetaminophen. The oil reduced lipid peroxidation and reactive oxygen species levels in the liver. Collectively, this suggests that lemongrass oil protects the liver against harmful substances generated during drug metabolism.

In vitro research suggests that melissa oil may prevent and manage neurodegenerative diseases caused by oxidative stress by inhibiting acetylcholinesterase (AChE) and butyrylcholinesterase (BChE) enzyme activity, and preventing lipid peroxidation.[510,511] Based on this, melissa oil shows potential to benefit the brain by boosting acetylcholine levels (which supports cognitive function) and by providing antioxidant protection against damaging oxidative stress, both of which are critical in the context of neurodegenerative diseases.

Hypothyroidism—a condition where the thyroid gland is underactive and doesn't produce enough thyroid hormones—can cause oxidative stress, which can damage tissues throughout the body. Administration of myrtle essential oil improved antioxidant capacity in rats with hypothyroidism.[512] Myrtle is a promising oil for protection against systemic oxidative stress associated with hypothyroidism, thereby supporting overall health and potentially alleviating some of the secondary complications of the condition.

Gastric (or stomach) ulcers are common in modern society and caused by a variety of factors including *H. pylori* infection, NSAIDs use, psychological stress, heavy alcohol consumption, and smoking. Evidence suggests that oxidative stress plays a significant role in the damage of the stomach lining. The protective effect of myrtle CT myrtenyl acetate essential oil was evaluated in rats with ulcers caused by ethanol/hydrogen chloride.[513] Rats were divided into six groups: 1) normal control, 2) ulcer control, 3) 250 mg/kg body weight myrtle oil, 4) 500 mg/kg BW myrtle oil, 5) 1,000 mg/kg BW myrtle oil, and 6) famotidine (brand name Pepcid—a stomach acid lowering medication used to treat ulcers and GERD). Oral administration of microencapsulated myrtle essential oil before exposure to ethanol/hydrogen chloride significantly protected against gastric ulcers, with the 500 and 1,000 mg dosage groups performing better than famotidine. The protective benefits of the oil were attributed to a balancing effect in overall stomach acidity, potent anti-inflammatory effects, prevention of the depletion of antioxidant enzyme activity (SOD, CAT, GPx), and reduced gastric lipoperoxidation—a chain reaction that oxidizes fats, primarily polyunsaturated fatty acids in cell membranes. Overall, the research suggests that oral use of myrtle essential oil reduces oxidative stress and inflammation that contribute to gastric ulcers.

Zebrafish (*Danio rerio*) are commonly used in medical and health research because they share about 70% of their genes with humans, and over 80% of genes linked to human diseases have a similar version in zebrafish. One of their biggest advantages is that their embryos are transparent, allowing scientists to easily observe how organs and tissues develop and how diseases progress—all without surgery. Zebrafish also grow and develop quickly, with major organs forming within just two to three days, which helps speed up research. They reproduce in large numbers, making it easy to gather enough data for reliable results. Researchers can easily modify their genes using modern tools like CRISPR. In addition, zebrafish show clear behaviors that allow for studies on brain function and mental health. They are used to study a wide range of human diseases, including cancer, heart disease, diabetes, and neurological conditions like Parkinson's and Alzheimer's. Exposure to oregano (*Origanum vulgare* subsp. *hirtum*, CT thymol) reduced anxiety-like behavior and improved cognition and brain oxidative stress, and decreased acetylcholinesterase activity in a zebrafish model.[514] Specifically, the oil reduced the negative effects of scopolamine by preserving SOD, CAT, GPx activity and GSH levels, while concurrently reducing malondialdehyde. In simple terms, the oil helped protect the brain from damage that can interfere with memory and thinking.

The effects of peppermint essential oil inhalation was evaluated in an APP/PS1 mouse model of Alzheimer's disease (a model that involves elevated beta-amyloid production associated with behavioral modifications and age-related deterioration of learning and memory).[515] Mice were split into three groups: 1) normal mice without AD, 2) mice with AD that inhaled peppermint oil, one hour, twice daily, for 21 days, and 3) mice with AD that inhaled rosemary oil, one hour, twice daily, for 21 days. Inhaling peppermint improved memory and the health of neurons in the hippocampus (returned to normal), and the deposition of amyloid reduced. In addition, markers of brain oxidative stress reduced, and enzymatic antioxidant activity (SOD, GPx) increased significantly to normal levels. Models were used to identify the mechanism of action and the researchers reported that peppermint improves cognitive function by improving amino acid metabolism (arginine, proline, cysteine, and methionine) and energy production, reducing brain oxidative damage, and protecting brain cells.

Sepsis is a potentially life-threatening condition that occurs when the body releases excess chemicals in response to an infection, which can damage multiple organ systems. Oral administration of deuterium-depleted water with rose essential oil (100 mg/kg body weight plus 15 to 30 PPM deuterium-depleted water) for two weeks reduced sepsis by regulating oxidative stress.[516] The natural solution reduced lipid peroxidation and inflammation, increased antioxidant activity, and healed liver tissue injuries caused by sepsis. Further research by the same authors concluded that rose essential oil alone (50 and 100 mg/kg BW) reduces oxidative injury caused by sepsis by reducing inflammation, increasing antioxidant activity, and protecting the liver.[517]

Like lavender, rosemary essential oil has been widely investigated for its role in mitigating the damage caused by oxidative stress. Animal research concluded that rosemary oil protects the male reproductive system against oxidative stress and functional damage in diabetic rats.[518] Rosemary essential oil (1,8-cineole CT) corrected depletions in antioxidant enzyme activities, oxidative stress, kidney and liver dysfunction, abnormal lipid profiles, and hyperglycemia in an experimental model of diabetes.[519] Oral administration of fennel or rosemary oils (or their respective nanoforms) in dyslipidemic rats improved biochemical parameters (reduced cardiovascular disease risk, decreased inflammation and oxidative stress, preserved liver function, and lowered blood sugar levels).[520] Rosemary oil increases intracellular reduced glutathione (GSH) levels and Nrf2-antioxidant responsive element (ARE)-reporter activity, which protects liver cells from damage and enhances liver cell survival *in vitro*.[521] Nrf2 is a potent protein found within all cells (but particularly in the liver and kidneys) that acts as the master regulator of the cellular antioxidant system when activated by AREs. Once triggered, Nrf2 activates over 200 genes that metabolize drugs and toxins, protect against oxidative stress, remove damaged proteins, and normalize inflammation. It also interacts with other cells that together determine longevity and protect against age-related diseases like cancer and neurological disorders. Rosemary essential oil (balanced CT) reversed anxiety, depression, memory impairment, and brain oxidative stress caused by scopolamine in zebrafish by reducing acetylcholinesterase (AChE) activity.[522] Collectively, these findings underscore rosemary essential oil's multifaceted potential as a natural therapeutic agent in safeguarding reproductive health, combating oxidative stress-related diseases, and supporting neuroprotection, thereby highlighting its promising role in enhancing overall well-being and longevity.

A thymol-rich form of thyme essential oil improved memory problems caused by scopolamine in zebrafish. The oil was administered by immersion once daily for thirteen days.[523] Thyme oil reduced scopolamine-induced increases in acetylcholinesterase (AChE) activity and the oxidative stress response, restoring brain antioxidant capacity. Oral administration of thyme essential oil as a preventive measure showed that it protects the liver during renal ischemia-reperfusion injuries by decreasing inflammation and oxidative stress.[524] Thyme oil is a promising natural solution for brain and liver health based on these findings.

In conclusion, while clinical data on the protective effects of essential oils against oxidative stress remains limited, extensive laboratory and animal research underscores their significant potential in combating oxidative stress and its associated health issues. Many essential oils demonstrate remarkable capabilities to neutralize free radicals, reduce inflammation, and bolster the body's natural antioxidant defenses through various mechanisms. This body of evidence highlights the promising role of essential oils as complementary agents in the prevention and management of conditions linked to oxidative damage, such as neurodegenerative diseases, cardiovascular disorders, and inflammatory ailments. Future research, including well-designed human clinical trials, is essential to fully understand their efficacy, optimal usage and dosage, and safety profiles at therapeutic levels. As we continue to explore the therapeutic potential of these natural compounds, essential oils will become more valuable in promoting overall health and resilience against oxidative stress-related diseases.

CHAPTER EIGHT

Embracing a Holistic Approach to Beat Oxidative Stress Naturally

As we reach the culmination of this comprehensive exploration into oxidative stress and its profound impact on our health, it is vital to recognize that addressing this complex biological challenge requires a multifaceted, holistic approach. Throughout this book, we have journeyed through the intricate mechanisms of oxidative stress, the insidious role of environmental and lifestyle factors, and the promising strategies rooted in nature to bolster our defenses. Now, as we synthesize these insights, it's up to you to incorporate them into a lifestyle that truly helps you bounce back, feel alive, and thrive over time with greater resilience, vitality, and long-term well-being.

Understanding the Interconnected Nature of Oxidative Stress

From the foundational principles outlined in Chapter 1, we understand that oxidative stress results from an imbalance between free radical production and the body's ability to neutralize them. This imbalance is not merely a biochemical anomaly, but a dynamic interplay influenced by myriad factors— from environmental exposures to our daily habits, emotions, and even genetics. Recognizing this interconnectedness is the first step toward meaningful intervention. It shifts the perspective from reactive treatment to proactive prevention, emphasizing the importance of lifestyle choices and environmental awareness.

Hidden Dangers and the Power of Awareness

Chapter 2 illuminated how electromagnetic fields (EMFs), often overlooked, can subtly increase oxidative load. While the modern world offers unparalleled technological advancements, awareness of these hidden dangers empowers us to make informed choices—such as reducing unnecessary device exposure, using shielding technologies, or optimizing our environment. Knowledge becomes a tool to mitigate risks that, over time, can contribute significantly to our oxidative burden.

Health Consequences and the Urgency for Action

The damaging health effects detailed in Chapter 3—ranging from neurodegenerative diseases to cardiovascular conditions, cancers, and aging—highlight the critical need for effective, natural countermeasures. These effects underscore that oxidative stress is not a mere laboratory concept but a real-world challenge with tangible, often debilitating consequences. This awareness should galvanize us to take proactive steps, knowing that the choices we make today influence our health trajectory tomorrow.

Lifestyle: The Foundation of Natural Defense

Chapter 4 emphasized that lifestyle factors are the cornerstone of managing oxidative stress. Nutrition, physical activity, sleep, hydration, toxin avoidance, and mental and emotional health are not isolated elements but parts of an interconnected system. For example, a nutrient-rich diet provides substrates for antioxidant defenses, while adequate sleep and stress management reduce endogenous free radical production. Regular exercise, when balanced, enhances mitochondrial function and antioxidant capacity. Conversely, toxin exposure and emotional distress exacerbate oxidative damage, creating a vicious cycle. Cultivating awareness and intentionality in these areas can dramatically tilt the balance in our favor.

Harnessing Dietary Antioxidants and Nutraceuticals

In Chapter 5, our focus was on the vital importance of obtaining antioxidants through whole, fresh foods—both plant-based and animal-derived—that naturally supply a broad spectrum of protective compounds. A diet rich in colorful fruits and vegetables, such as berries, apples, peppers, and root vegetables, provides an abundant source of vitamins, minerals, and polyphenols that combat oxidative stress at their source. Incorporating nuts, seeds, whole grains, and legumes adds essential nutrients like selenium, magnesium, and fiber, which support the body's internal defenses. Including high-quality, minimally processed animal products like eggs, fish, and grass-fed beef can also contribute vital nutrients such as coenzyme Q10 and zinc, which play roles in antioxidant activity. The key message is that a diverse, fresh, and minimally processed diet creates a powerful, natural arsenal against free radicals. While supplements can help address specific deficiencies, they should always serve as complements to, rather than substitutes for, a balanced diet rich in whole foods. The strongest evidence indicates that these natural antioxidants are most effective when they are part of an overall health-conscious lifestyle, emphasizing variety, freshness, and mindful eating habits. By prioritizing whole, nutrient-dense foods, you lay a sustainable foundation for reducing oxidative stress and supporting your long-term health.

The Power of Targeted Nutraceuticals

In addition to lifestyle modifications and natural remedies, Chapter 6 highlighted how targeted supplementation can play a crucial role in strengthening the body's defenses against oxidative stress. Key supplements such as alpha-lipoic acid, vitamin C, CoQ10, resveratrol, selenium, and quercetin have demonstrated potent antioxidant properties in scientific studies. Alpha-lipoic acid, often called the universal antioxidant, can regenerate other antioxidants and penetrate cell membranes to reduce intracellular oxidative damage. Vitamin C is a powerful water-soluble antioxidant that neutralizes free radicals and supports immune function. CoQ10 is essential for mitochondrial energy production and helps protect cells from oxidative injury, particularly in the heart and brain. Resveratrol, found in grapes and berries, activates cellular pathways that enhance antioxidant defenses and promote longevity. Selenium, a trace mineral, is a vital component of antioxidant enzymes like glutathione peroxidase, helping to detoxify peroxides. Quercetin, a flavonoid present in many fruits and vegetables, reduces inflammation and scavenges a wide range of free radicals. Incorporating these supplements, and others outlined previously, can provide an extra layer of protection, especially for individuals facing heightened oxidative challenges due to environmental exposures, aging, or health conditions. When combined with a balanced diet and healthy lifestyle, these targeted nutrients can significantly bolster your body's resilience and support overall wellness.

The Emerging Role of Essential Oils

Chapter 7 showcased the remarkable antioxidant potential of essential oils derived from nature's pharmacy, and their unique contribution to the human antioxidant defense system. Their ability to neutralize free radicals, reduce inflammation, and support cellular health adds a valuable dimension to natural strategies against oxidative stress. Whether used aromatically, topically, or ingested, essential oils can serve as potent natural solutions, aiding body systems in their pursuit to maintain overall health. Their incorporation requires mindful application and awareness of quality, but their inclusion can enrich a holistic health regimen.

Moving Toward a Personal and Community-Centered Lifestyle

While individual efforts are vital, fostering community awareness and support amplifies the impact. Sharing knowledge about environmental toxins, advocating for cleaner air and water, and promoting healthy lifestyle choices can create ripple effects that benefit not only ourselves but also the communities around us, and future generations. Supporting policies that reduce environmental pollutants and exposure to EMFs aligns with the principle that collective action can shape healthier environments.

Mindfulness, Emotions, and Inner Balance

An often-overlooked aspect of oxidative stress management is emotional health. Chronic stress, anxiety, and unresolved emotional conflicts can elevate oxidative markers. Cultivating mindfulness, meditation, deep breathing, and nurturing social connections can soothe the nervous system, lowering endogenous free radical production. Inner harmony and emotional resilience are integral to maintaining oxidative balance.

A Sustainable, Personalized Approach

Every individual's genetic makeup, biological diversity, environment, lifestyle, and health status vary. Therefore, personalized strategies—guided by informed healthcare professionals—are essential. Regular assessment of oxidative markers, nutritional status, and environmental exposures can inform tailored interventions. Combining dietary choices, lifestyle modifications, natural supplements, and mindful practices creates a resilient, adaptable defense system.

The Power of Consistency and Patience

Combatting oxidative stress is not a one-time effort but a lifelong commitment. Small, consistent daily choices—such as eating healthy meals, engaging in regular movement, practicing mindfulness, or diffusing essential oils—compound over time. Patience and perseverance are key, as biological systems adapt gradually, and cumulative effects become apparent.

Looking Ahead: The Future of Natural Oxidative Stress Management

Advances in research continue to uncover novel plant compounds, bioactive molecules, and integrative therapies that support our natural defenses. Emerging technologies for environmental monitoring and personalized medicine will further empower us to optimize our internal and external environments. Embracing a mindset of curiosity and ongoing learning ensures we stay informed and adaptable in our health journeys.

Now Is the Time to Act

Finally, you are invited to view this knowledge not merely as information but as a call to empower yourself. Small, conscious changes—rooted in understanding and respect for nature's wisdom—can profoundly influence your health. Embrace a lifestyle that honors your body's innate capacity to heal and protect itself. Cultivate awareness of your environment, nourish your body with vibrant foods and natural remedies, nurture your emotional well-being, and foster community connections.

Remember, combating oxidative stress naturally is an ongoing journey of self-care, mindfulness, and environmental stewardship. By integrating the insights from this book into your daily life, you can enhance your resilience, slow the aging processes, and enjoy a healthier, more vibrant existence. Your body is your most precious gift—treat it with love, respect, and the wisdom of nature.

Together, through informed choices and compassionate living, we can turn the tide against oxidative stress and nurture a healthier planet and population.

References

[1] Hahm JY, Park J, Jang ES, et al. 8-Oxoguanine: from oxidative damage to epigenetic and epitranscriptional modification. *Exp Molecular Med*. 2022;54:1626-42.

[2] Singh SK, Szulik MW, Ganguly M, et al. Characterization of DNA with an 8-oxoguanine modification. *Nucleic Acids Res*. 2011 Aug;39(15):6789-801.

[3] Schumacher B, Pothof J, Vijg J, et al. The central role of DNA damage in the ageing process. *Nature*. 2021;592:695-703.

[4] Lagouge M, Larsson MG. The role of mitochondrial DNA mutations and free radicals in disease and ageing. *J Intern Med*. 2013 Mar 7;273(6):529–543.

[5] Hayes JD, Dinkova-Kostova AT, Tew KD. Oxidative Stress in Cancer. *Cancer Cell*. 2020 Jul 9;38(2):167–197.

[6] Arfin S, Jha NK, Jha SK, et al. Oxidative Stress in Cancer Cell Metabolism. *Antioxidants (Basel)*. 2021 Apr 22;10(5):642.

[7] Anjo S, He Z, Hussain Z, et al. Protein Oxidative Modifications in Neurodegenerative Diseases: From Advances in Detection and Modelling to Their Use as Disease Biomarkers. *Antioxidants (Basel)*. 2024 May 31;13(6):681.

[8] Kellar D, Craft S. Brain insulin resistance in Alzheimer's disease and related disorders: mechanisms and therapeutic approaches. *Lancet Neurol*. 2020 Sep;19(9):758-766.

[9] Holscher C. Brain insulin resistance: role in neurodegenerative disease and potential for targeting. *Expert Opin Investig Drugs*. 2020 Apr;29(4):333-348.

[10] Dash UC, Bhal NK, Swain SK, et al. Oxidative stress and inflammation in the pathogenesis of neurological disorders: Mechanisms and implications. *Acta Pharma Sinica B*. 2025 Jan;15(1):15-34.

[11] Dubois-Deury E, Peugnet V, Turkieh A, et al. Oxidative Stress in Cardiovascular Diseases. *Antioxidants (Basel)*. 2020 Sep 14;9(9):864.

[12] Pizzino G, Irrera N, Cucinotta M, et al. Oxidative Stress: Harms and Benefits for Human Health. *Oxid Med Cell Longev*. 2017 Jul 27;2017:8416763.

[13] Miller AB, Sears ME, Morgan LL, et al. Risks to Health and Well-Being From Radio-Frequency Radiation Emitted by Cell Phones and Other Wireless Devices. *Front Public Health*. 2019 Aug 13;7:223.

[14] Schuermann D, Mevissen M. Manmade Electromagnetic Fields and Oxidative Stress—Biological Effects and Consequences for Health. *Int J Mol Sci*. 2021 Apr 6;22(7):3772.

[15] Duan W, Liu C, Zhang L, et al. Comparison of the genotoxic effects induced by 50 Hz extremely low-frequency electromagnetic fields and 1800 MHz radiofrequency electromagnetic fields in GC-2 cells. *Radiat Res*. 2015 Mar;183(3):305-14.

[16] Liu Y, Liu WB, Liu KJ, et al. Effect of 50 Hz Extremely Low-Frequency Electromagnetic Fields on the DNA Methylation and DNA Methyltransferases in Mouse Spermatocyte-Derived Cell Line GC-2. *Biomed Res Int*. 2015:2015:237183.

[17] Liu Y, Liu WB, Liu KJ, et al. Overexpression of miR-26b-5p regulates the cell cycle by targeting CCND2 in GC-2 cells under exposure to extremely low frequency electromagnetic fields. *Cell Cycle*. 2016;15(3):357-67.

[18] Kivrak EG, Yurt KK, Kaplan AA, et al. Effects of electromagnetic fields exposure on the antioxidant defense system. *J Microscopy Ultrastructure*. 2017;5:167-76.

[19] Kivrak EG, Yurt KK, Kaplan AA, et al. Effects of electromagnetic fields exposure on the antioxidant defense system. *J Microscopy Ultrastructure*. 2017;5:167-76.

[20] Kivrak EG, Yurt KK, Kaplan AA, et al. Effects of electromagnetic fields exposure on the antioxidant defense system. *J Microsc Ultrastruct*. 2017 Aug 2;5(4):167–176.

[21] Sahu M, Behera S, Chattopadhyay B. The Influence of Electromagnetic Field Pollution on Human Health: A Systematic Review. *Siriraj Med J*. 2021 Jul;73(7):485-92.

[22] Terzi M, Ozberk B, Deniz OG, et al. The role of electromagnetic fields in neurological disorders. *J Chem Neuroanat*. 2016 Sep;75(Pt B):77-84.

[23] Schuermann D, Mevissen M. Manmade Electromagnetic Fields and Oxidative Stress— Biological Effects and Consequences for Health. *Int J Mol Sci*. 2021;22:3772.

[24] Georgiou CD, Margaritis LH. Oxidative Stress and NADPH Oxidase: Connecting Electromagnetic Fields, Cation Channels and Biological Effects. *Int J Mol Sci*. 2021;22:10041.

[25] Santini SJ, Cordone V, Falone S, et al. Role of Mitochondria in the Oxidative Stress Induced by Electromagnetic Fields: Focus on Reproductive Systems. *Oxid Med Cellular Longevity*. 2018;2018:5076271.

[26] Martinez-Samano J, Torres-Duran PV, Juarez-Oropeza MA, et al. Effects of acute electromagnetic field exposure and movement restraint on antioxidant system in liver, heart, kidney and plasma of Wistar rats: a preliminary report. *Int J Radiat Biol*. 2010 Dec;86(12):1088-94.

[27] Ghlampour F, Owji SM, Javadifar TS. Chronic Exposure to Extremely Low Frequency Electromagnetic Field Induces Mild Renal Damages in Rats. *Int J Zoological Res*. 2011;7(6):393-400.

[28] Borzoueisileh S, Monfared AS, Ghorbani H, et al. Assessment of function, histopathological changes, and oxidative stress in liver tissue due to ionizing and non-ionizing radiations. *Caspian J Intern Med*. 2020 May;11(3):315–323.

[29] INTERPHONE Study Group. Brain tumour risk in relation to mobile telephone use: results of the INTERPHONE international case-control study. *Int J Epidemiol*. 2010 Jun;39(3):675-94.

[30] Kundi M, Mild K, Hardell L, et al. Mobile telephones and cancer--a review of epidemiological evidence. *J Toxicol Environ Health B Crit Rev*. 2004 Sep-Oct;7(5):351-84.

[31] Choi YJ, Moskowitz JM, Myung SK, et al. Cellular Phone Use and Risk of Tumors: Systematic Review and Meta-Analysis. *Int J Environ Res Public Health*. 2020 Nov 2;17(21):8079.

[32] Zhao L, Liu X, Wang C, et al. Magnetic fields exposure and childhood leukemia risk: a meta-analysis based on 11,699 cases and 13,194 controls. *Leuk Res*. 2014 Mar;38(3):269-74.

[33] Nunez-Enriquez JC, Correa-Correa V, Flores-Lujano J, et al. Extremely Low-Frequency Magnetic Fields and the Risk of Childhood B-Lineage Acute Lymphoblastic Leukemia in a City With High Incidence of Leukemia and Elevated Exposure to ELF Magnetic Fields. *Bioelectromagnetics*. 2020 Dec;41(8):581-597.

[34] Brabant C, Geerick A, Beaudart C, et al. Exposure to magnetic fields and childhood leukemia: a systematic review and meta-analysis of case-control and cohort studies. *Rev Environ Health*. 2022 Mar 15;38(2):229-253.

[35] Liu H, Chen G, Pan Y, et al. Occupational Electromagnetic Field Exposures Associated with Sleep Quality: A Cross-Sectional Study. *PLoS One*. 2014 Oct 23;9(10):e110825.

[36] Hosseinabadi MB, Khanjani N, Ebrahimi MH, et al. The effect of chronic exposure to extremely low-frequency electromagnetic fields on sleep quality, stress, depression and anxiety. *Electromagn Biol Med*. 2019;38(1):96-101.

[37] Mohler E, Frei P, Frohlich J, et al. Exposure to Radiofrequency Electromagnetic Fields and Sleep Quality: A Prospective Cohort Study. *PLoS One*. 2012 May 18;7(5):e37455.

[38] Danker-Hopfe H, Bueno-Lopez A, Dorn H, et al. Spending the night next to a router - Results from the first human experimental study investigating the impact of Wi-Fi exposure on sleep. *Int J Hyg Environ Health*. 2020 Jul:228:113550.

[39] Dieudonne M. Electromagnetic hypersensitivity: a critical review of explanatory hypotheses. *Environ Health*. 2020 May 6;19:48.

[40] Belpomme D, Irigaray P. Electrohypersensitivity as a Newly Identified and Characterized Neurologic Pathological Disorder: How to Diagnose, Treat, and Prevent It. *Int J Mol Sci*. 2020 Mar 11;21(6):1915.

[41] Irgaray P, Caccamo D, Belpomme D. Oxidative stress in electrohypersensitivity self-reporting patients: Results of a prospective in vivo investigation with comprehensive molecular analysis. *Int J Mol Med*. 2018 Oct;42(4):1885-1898

[42] Belpomme D, Campagnac C, Irigaray P. Reliable disease biomarkers characterizing and identifying electrohypersensitivity and multiple chemical sensitivity as two etiopathogenic aspects of a unique pathological disorder. *Rev Environ Health*. 2015;30(4):251-71.

[43] World Health Organization. Electromagnetic fields. Accessed May 22, 2025 from https://www.who.int/india/health-topics/electromagnetic-fields.

[44] Chou CK. Controversy in Electromagnetic Safety. *Int J Environ Res Public Health*. 2022 Dec 16;19(24):16942.

[45] Maldonado E, Morales-Pison S, Urbina F, et al. Aging Hallmarks and the Role of Oxidative Stress. *Antioxidants*. 2023;12(3):651.

[46] Gorbunova V, Seluanov A, Mao Z, et al. Changes in DNA repair during aging. *Nucleic Acids Res*. 2007 Oct 2;35(22):7466–7474.

[47] Gambini J, Stromsnes K. Oxidative Stress and Inflammation: From Mechanisms to Therapeutic Approaches. *Biomedicines*. 2022 Mar 23;10(4):753.

[48] Shen C, Jiang X, Luo Y. Hepatitis Virus and Hepatocellular Carcinoma: Recent Advances. *Cancers (Basel)*. 2023 Jan 15;15(2):533.

[49] Zareba K, Dorf J, Cummings K, et al. Associations of oxidative stress, metabolic disorders in colorectal cancer patients. *Prz Gastroenterol*. 2024 May 28;19(2):206–213.

[50] Hadi NA, Reyes-Castellanos G, Carrier A. Targeting Redox Metabolism in Pancreatic Cancer. *Int J Mol Sci*. 2021 Feb 3;22(4):1534.

[51] An X, Yu W, Liu J, et al. Oxidative cell death in cancer: mechanisms and therapeutic opportunities. *Cell Death Dis*. 2024;15:556 (2024).

[52] Rains JL, Jain SK. OXIDATIVE STRESS, INSULIN SIGNALING AND DIABETES. *Free Radic Biol Med*. 2010 Dec 13;50(5):567–575.

[53] Eguchi N, Vaziri ND, Dafoe DC, et al. The Role of Oxidative Stress in Pancreatic β Cell Dysfunction in Diabetes. *Int J Mol Sci*. 2021 Feb 3;22(4):1509.

[54] Giacco F, Brownlee M. Oxidative stress and diabetic complications. *Circ Res*. 2010 Oct 29;107(9):1058-70.

[55] Martemucci G, Fracchiolla G, Muraglia M, et al. Metabolic Syndrome: A Narrative Review from the Oxidative Stress to the Management of Related Diseases. *Antioxidants (Basel)*. 2023 Dec 8;12(12):2091.

[56] Masenga SK, Kabwe LS, Chakulya M, et al. Mechanisms of Oxidative Stress in Metabolic Syndrome. *Int J Mol Sci*. 2023 Apr 26;24(9):7898.

[57] Incalza MA, D'Oria R, Natalicchio A, et al. Oxidative stress and reactive oxygen species in endothelial dysfunction associated with cardiovascular and metabolic diseases. *Vascul Pharmacol*. 2018 Jan:100:1-19.

[58] Higasji Y. Roles of Oxidative Stress and Inflammation in Vascular Endothelial Dysfunction-Related Disease. *Antioxidants (Basel)*. 2022 Sep 30;11(10):1958.

[59] Cheignon C, Tomas M, Bonnefont-Rousselot D, et al. Oxidative stress and the amyloid beta peptide in Alzheimer's disease. *Redox Biol*. 2017 Oct 18;14:450–464.

[60] Deng Y, Li B, Liu Y, et al. Dysregulation of Insulin Signaling, Glucose Transporters, O-GlcNAcylation, and Phosphorylation of Tau and Neurofilaments in the Brain. *Am J Pathol*. 2009 Nov;175(5):2089–2098.

[61] Abolhassani N, Leon J, Sheng Z, et la. Molecular pathophysiology of impaired glucose metabolism, mitochondrial dysfunction, and oxidative DNA damage in Alzheimer's disease brain. *Mech Ageing Dev*. 2017 Jan;161(Pt A):95-104.

[62] Rhea EM, Banks WA, Raber J. Insulin Resistance in Peripheral Tissues and the Brain: A Tale of Two Sites. *Biomedicines*. 2022 Jul 2;10(7):1582.

[63] Blazquez E, Hurtado-Carneiro V, LeBaut-Ayuso Y, et al. Significance of Brain Glucose Hypometabolism, Altered Insulin Signal Transduction, and Insulin Resistance in Several Neurological Diseases. *Front Endocrinol (Lausanne)*. 2022 May 9;13:873301.

[64] Guo JD, Zhao X, Li Y, et al. Damage to dopaminergic neurons by oxidative stress in Parkinson's disease (Review). *Int J Mol Med*. 2018 Apr;41(4):1817-1825.

[65] Sian-Hulsmann J, Riederer P, Michel TM/ Metabolic Dysfunction in Parkinson's Disease: Unraveling the Glucose–Lipid Connection. *Biomedicines*. 2024 Dec 13;12(12):2841.

[66] Allameh A, Niayesh-Mehr R, Aliarab A, et al. Oxidative Stress in Liver Pathophysiology and Disease. *Antioxidants (Basel)*. 2023 Aug 22;12(9):1653.

[67] Gyuraszova M, Gurecka R, Babuckova J, et al. Oxidative Stress in the Pathophysiology of Kidney Disease: Implications for Noninvasive Monitoring and Identification of Biomarkers. *Oxid Med Cell Longev*. 2020 Jan 23;2020:5478708.

[68] Assavarittirong C, Samborksi W, Grygiel-Gorniak B. Oxidative Stress in Fibromyalgia: From Pathology to Treatment. *Oxid Med Cell Longev*. 2022 Oct 5;2022:1582432.

[69] Cordero MD, Alcoer-Gomez E, Cano-Garcia FJ, et al. Clinical Symptoms in Fibromyalgia Are Better Associated to Lipid Peroxidation Levels in Blood Mononuclear Cells Rather than in Plasma. *PLoS One*. 2011 Oct 28;6(10):e26915.

[70] Cordero MD, De Miguel M, Fernandez AMM, et al. Mitochondrial dysfunction and mitophagy activation in blood mononuclear cells of fibromyalgia patients: implications in the pathogenesis of the disease. *Arthritis Res Ther*. 2010 Jan 28;12(1):R17.

[71] Lee JS, Kim HG, Lee DS, et al. Oxidative Stress is a Convincing Contributor to Idiopathic Chronic Fatigue. *Sci Rep*. 2018 Aug 27;8:12890.

[72] Kennedy G, Spence VA, McLaren M, et al. Oxidative stress levels are raised in chronic fatigue syndrome and are associated with clinical symptoms. *Free Radic Biol Med*. 2005 Sep 1;39(5):584-9.

[73] Maes M, Mihaylova I, De Ruyter M. Lower serum zinc in Chronic Fatigue Syndrome (CFS): Relationships to immune dysfunctions and relevance for the oxidative stress status in CFS. *J Affective Disorders*. 2006 Feb;90(2-3):141-7.

[74] Yilmaz MI, Romano M, Basatali MK, et al. The Effect of Corrected Inflammation, Oxidative Stress and Endothelial Dysfunction on Fmd Levels in Patients with Selected Chronic Diseases: A Quasi-Experimental Study. *Sci Reports*. 2020;10(9018):2020.

[75] Bhol NK, Bhanjadeo MM, Singh AK, et al. The interplay between cytokines, inflammation, and antioxidants: mechanistic insights and therapeutic potentials of various antioxidants and anti-cytokine compounds. *Biomed Pharmacother*. 2024 Sep;178:117177.

[76] Solmaz D, Koxaci D, Sari I, et al. Oxidative stress and related factors in patients with ankylosing spondylitis. *Eur J Rheumatol*. 2016 Jan 29;3(1):20–24.

[77] Ye G, Xie Z, Zeng H, et al. Oxidative stress-mediated mitochondrial dysfunction facilitates mesenchymal stem cell senescence in ankylosing spondylitis. *Cell Death Dis*. 2020;11(775):2020.

[78] Liu Y, Chen Q. Senescent Mesenchymal Stem Cells: Disease Mechanism and Treatment Strategy. *Curr Mol Biol Rep*. 2020 Oct 28;6(4):173–182.

[79] Chainy GBN, Sahoo DK. Hormones and oxidative stress: an overview. *Free Radic Res*. 2020 Jan;54(1):1-26.

[80] Ide T, Tsutsi H, Ohashi N, et al. Greater oxidative stress in healthy young men compared with premenopausal women. *Arterioscler Thromb Vasc Biol*. 2002;22:438-42.

[81] Lavoie JC, Tremblay A. Sex-specificity of oxidative stress in newborns leading to a personalized antioxidant nutritive strategy. *Antioxidants (Basel)*. 2018 Mar 27;7(4):49.

[82] Mair KM, Gaw R, MacLean MR. Obesity, estrogens and adipose tissue dysfunction – implications for pulmonary arterial hypertension. *Pulm Circ*. 2020 Sep 18;10(3):2045894020952019.

[83] Liu J, Ma C., Jeng Y, et al. Association between the oxidative balance score and testosterone deficiency: a cross-sectional study of the NHANES, 2011–2016. *Sci Reports.* 2025;15(8040):2025.

[84] Maharjan CK, Mo J, Wang L, et al. Natural and Synthetic Estrogens in Chronic Inflammation and Breast Cancer. *Cancers (Basel).* 2021 Dec 31;14(1):206.

[85] Killinger DW, Perel E, Daniilescu D, et al. The relationship between aromatase activity and body fat distribution. *Steroids.* 1987 Jul-Sep;50(1-3):61-72.

[86] Cohen PG. Obesity in men: the hypogonadal-estrogen receptor relationship and its effect on glucose homeostasis. *Med Hypotheses.* 2008;70(2):358-60.

[87] Bellanti F, Matteo M, Rollo T, et al. Sex hormones modulate circulating antioxidant enzymes: impact of estrogen therapy. *Redox Biol.* 2013;1:340-346.

[88] Suojanen JN, Gay RJ, Hilf R. Influence of estrogen on glutathione levels and glutathione-metabolizing enzymes in uteri and R3230AC mammary tumors of rats. *Biochim Biophys Acta.* 1980;630:485-496.

[89] Borras C, Gambini J, Lopez-Grueso R, et al. Direct antioxidant and protective effect of estradiol on isolated mitochondria. *Biochim Biophys Acta.* 2010;1802:205-211.

[90] Brieger K, Schiavone S, Miller Jr. FJ, et al. Reactive oxygen species: from health to disease. *Swiss Med Wkly.* 2012 Aug 17:142:w13659.

[91] Kunsch C, Medford RM. Oxidative Stress as a Regulator of Gene Expression in the Vasculature. *Circulation Res.* 1999 Oct 15;85(8):753-66.

[92] Mohammadi M. Oxidative Stress and Polycystic Ovary Syndrome: A Brief Review. *Int J Prev Med.* 2019 May 17;10:86.

[93] Zaha I, Murasan M, Tulcan C, et al. The Role of Oxidative Stress in Infertility. *J Pers Med.* 2023 Aug 15;13(8):1264.

[94] Rajimakers MTM, Dechend R, Poston L. Oxidative Stress and Preeclampsia: Rationale for Antioxidant Clinical Trials. *Hypertension.* 2004 Aug 23;44(4):374-80.

[95] Papaccio F, D'Arino A, Caputo S, et al. Focus on the Contribution of Oxidative Stress in Skin Aging. *Antioxidants (Basel).* 2022 Jun 6;11(6):1121.

[96] Subash S, Essa MM, Al-Adawi S, et al. Neuroprotective effects of berry fruits on neurodegenerative diseases. *Neural Regen Res.* 2014 Aug 15;9(16):1557–1566.

[97] Devore EE, Kang JH, Breteler MMB, et al. Dietary intake of berries and flavonoids in relation to cognitive decline. *Ann Neurol.* 2012 Apr 26;72(1):135–143.

[98] Gallone G, Haerty W, Disanto G, et al. Identification of genetic variants affecting vitamin D receptor binding and associations with autoimmune disease. *Human Mol Genetics.* 2017 Jun 1;26(11):2164-76.

[99] Hossein-nezhad A, Spira A, Holicj MF. Influence of Vitamin D Status and Vitamin D3 Supplementation on Genome Wide Expression of White Blood Cells: A Randomized Double-Blind Clinical Trial. *PLoS One.* 2013 Mar 20;8(3):e58725.

[100] Pasing Y, Fenton CG, Jorde R, et al. Changes in the human transcriptome upon vitamin D supplementation. *J Steroid Biochem Mol Biol.* 2017 Oct:173:93-99.

[101] During A, Harrison EH. Mechanisms of provitamin A (carotenoid) and vitamin A (retinol) transport into and out of intestinal Caco-2 cells. *J Lipid Res.* 2007;48:2283–2294.

[102] Schreiber R, Taschler U, Preiss-Landl K, et al. Retinyl ester hydrolases and their roles in vitamin A homeostasis. *Biochim Biophys Acta.* 2012;1821:113–123.

[103] Tang G, Qin J, Dolnikowski GG, et al. Short-term (intestinal) and long-term (postintestinal) conversion of beta-carotene to retinol in adults as assessed by a stable-isotope reference method. *Am J Clin Nutr.* 2003 Aug;78(2):259-66.

[104] Far BF, Lomer NB, Gharedaghi H, et al. Is beta-carotene consumption associated with thyroid hormone levels? *Front Endocrinol (Lausanne).* 2023 May 26;14:1089315.

[105] Lieber CS. Relationships Between Nutrition, Alcohol Use, and Liver Disease. *Alcohol Res Health.* 2003;27(3):220–231.

[106] U.S. National Institutes of Health. Vitamin A and Carotenoids: Fact Sheet for Health Professionals. Available at: https://ods.od.nih.gov/factsheets/VitaminA-HealthProfessional/, accessed March 26, 2025.

[107] Leung WC, Hessel S, Meplan C, et al. Two common single nucleotide polymorphisms in the gene encoding beta-carotene 15,15'-monoxygenase alter beta-carotene metabolism in female volunteers. *FASEB J*. 2009 Apr;23(4):1041-53.

[108] Piskin E, Cianciosi D, Gulec S, et al. Iron Absorption: Factors, Limitations, and Improvement Methods. *ACS Omega*. 2022 Jun 10;7(24):20441–20456.

[109] Petroski W, Minich DM. Is There Such a Thing as "Anti-Nutrients"? A Narrative Review of Perceived Problematic Plant Compounds. *Nutrients*. 2020 Sep 24;12(10):2929.

[110] Vogiatzoglou A, Refsum H, Johnston C, et al. Vitamin B12 status and rate of brain volume loss in community-dwelling elderly. *Neurology*. 2008 Sep 9;71(11):826-32.

[111] Kocaoglu C, Akin F, Caksen H, et al. Cerebral atrophy in a vitamin B12-deficient infant of a vegetarian mother. *J Health Popul Nutr*. 2014 Jun;32(2):367-71.

[112] Mathey C, Di Marco JN, Poujol A, et al. [Failure to thrive and psychomotor regression revealing vitamin B12 deficiency in 3 infants]. *Arch Pediatr*. 2007 May;14(5):467-71.

[113] Bousselamti A, El Hasbaoui B, Echahdi H, et al. Psychomotor regression due to vitamin B12 deficiency. *Pan Afr Med J*. 2018 Jun 20:30:152.

[114] Smolka V, Bekarek V, Hlidkocva E, et al. [Metabolic complications and neurologic manifestations of vitamin B12 deficiency in children of vegetarian mothers]. *Cas Lek Cesk*. 2001 Nov 22;140(23):732-5.

[115] Supruniuk E, Gorski J, Chabowski A. Endogenous and Exogenous Antioxidants in Skeletal Muscle Fatigue Development during Exercise. *Antioxidants (Basel)*. 2023 Feb 16;12(2):501.

[116] de Sousa CV, Sales MM, Rosa TS, et al. The Antioxidant Effect of Exercise: A Systematic Review and Meta-Analysis. *Sports Med*. 2017 Feb;47(2):277-293.

[117] Bird SR, Hawley JA. Update on the effects of physical activity on insulin sensitivity in humans. *BMJ Open Sport Exerc Med*. 2017 Mar 1;2(1):e000143.

[118] Buresh R. Exercise and glucose control. *J Sports Med Phys Fitness*. 2014 Aug;54(4):373-82.

[119] Pagan LU, Gomes MJ, Okoshi MP. Endothelial Function and Physical Exercise. *Arq Bras Cardiol*. 2018 Oct;111(4):540–541.

[120] Mahindru A, Patil P, Agrawal V. Role of Physical Activity on Mental Health and Well-Being: A Review. *Cureus*. 2023 Jan 7;15(1):e33475.

[121] Noetel M, Sanders T, Gallardo-Gomez D, et al. Effect of exercise for depression: systematic review and network meta-analysis of randomised controlled trials. *BMJ*. 2024;384.

[122] Shamsnia E, Matinhomaee H, Azarbayjani MA, et al. The Effect of Aerobic Exercise on Oxidative Stress in Skeletal Muscle Tissue: A Narrative Review. *Gene Cell Tissue*. 2023 Oct;10(4):e131964.

[123] Hough J. Overtraining and the endocrine system: Can hormones indicate overtraining? *Endocrinologist*. 2024 Autumn;153:Autumn2024.

[124] Everson CA, Laatsch CD, Hogg N. Antioxidant defense responses to sleep loss and sleep recovery. *Am J Physiol Regul Integr Comp Physiol*. 2005;288:R374–R383.

[125] Richardson RB, Mailloux RJ. Mitochondria Need Their Sleep: Redox, Bioenergetics, and Temperature Regulation of Circadian Rhythms and the Role of Cysteine-Mediated Redox Signaling, Uncoupling Proteins, and Substrate Cycles. *Antioxidants (Basel)*. 2023 Mar 9;12(3):674.

[126] Atrooz F, Salim S. Sleep deprivation, oxidative stress and inflammation. *Adv Protein Chem Struct Biol*. 2020:119:309-336.

[127] Tang L, Liu M, Mu J, et al. Association between circulating antioxidants and sleep disorders: comprehensive results from NHANES 2017–2018. *Food Function*. 2024 Jun 17;15(12):6657-6672.

[128] Davinelli S, Medoro A, Savino R, et al. Sleep and Oxidative Stress: Current Perspectives on the Role of NRF2. *Cell Mol Neurobiol*. 2024 Jun 25;44:52.

[129] Al-Madboly LA, Ali SM, Faharany EM, et al. Stress-Based Production, and Characterization of Glutathione Peroxidase and Glutathione S-Transferase Enzymes From Lactobacillus plantarum. *Front Bioeng Biotechnol*. 2020 Feb 26;8:78.

[130] Iizuka M, Inoue Y, Murata K, et al. Purification and some properties of glutathione S-transferase from Escherichia coli B. *J Bacteriol*. 1989 Nov;171(11):6039–6042.

[131] Nowak K, Ratajczak-Wrona W, Gorska M, et al. Parabens and their effects on the endocrine system. *Mol Cell Endocrinol*. 2018 Oct 15:474:238-251.

[132] Perez-Diaz C, Perez-Carrascosa FM, Riquelme-Gallego B, et al. Serum Phthalate Concentrations and Biomarkers of Oxidative Stress in Adipose Tissue in a Spanish Adult Cohort. *Environ Sci Tech*. 2024 Apr 23;58(18):7719-30.

[133] Ungureanu LBm Chiciuc CM, Amalinei C, et al. Antioxidants as Protection against Reactive Oxygen Stress Induced by Formaldehyde (FA) Exposure: A Systematic Review. *Biomedicines*. 2024 Aug 10;12(8):1820.

[134] Weatherly LM, Gosse JA. Triclosan Exposure, Transformation, and Human Health Effects. *J Toxicol Environ Health B Crit Rev*. 2017;20(8):447–469.

[135] Salar-Flores J, Jasso JT, Rojas-Bravo D, et al. Effects of Mercury, Lead, Arsenic and Zinc to Human Renal Oxidative Stress and Functions: A Review. *J Heavy Metal Toxicity Dis*. 2019 Jan;04(01):1-16.

[136] National Library of Medicine. 15th Report on Carcinogens [Internet]. Coal Tars and Coal-Tar Pitches. Accessed May 20, 2025 at: https://www.ncbi.nlm.nih.gov/books/NBK590777/.

[137] Chen Y, Wang Y, Charkofataki G, et al. Oxidative stress and genotoxicity in 1,4-dioxane liver toxicity as evidenced in a mouse model of glutathione deficiency. *Sci The Total Environ*. 2022 Feb;806(Part 2):150703.

[138] Sus S, Pham C, Smith J, et al. The Banned Sunscreen Ingredients and Their Impact on Human Health: A Systematic Review. *Int J Dermatol*. 2020 Feb 28;59(9):1033–1042.

[139] Steinemann A. Fragranced consumer products: exposures and effects from emissions. *Air Qual Atmos Health*. 2016 Oct 20;9(8):861–866.

[140] Saleh AA, Albohomoud MS, Almojiwel TAM, et al. Ammonia Toxicity: Integrating Environmental Health, Radiology, Nursing, and Respiratory Therapy. *J Ecohumanism*. 2024 Dec;3(8):13903-14.

[141] National Library of Medicine. Toxicological Profile for Chlorine Dioxide and Chlorite. Accessed May 20, 2025 from https://www.ncbi.nlm.nih.gov/books/NBK596896/.

[142] Shi X, Wang Y, Yang F, et al. Associations of exposure to volatile organic compounds with biological aging: a cross-sectional study. *BMC Public Health*. 2025 Apr 22;25:1476.

[143] Chung I, Ryu H, Yoon SY, er al. Health effects of sodium hypochlorite: review of published case reports. *Environ Anal Health Toxicol*. 2022 Mar 30;37(1):e2022006.

[144] Galli FS, Mollari M, Tassinari V, et al. Overview of human health effects related to glyphosate exposure. *Front Toxicol*. 2024 Sep 18;6:1474792.

[145] Aschbacher K, O'Donovan A, Wolkowitz OM, et al. Good Stress, Bad Stress and Oxidative Stress: Insights from Anticipatory Cortisol Reactivity. *Psychoneuroendocrinology*. 2013 Mar 13;38(9):1698–1708.

[146] Kerahrodi JG, Michal M. The fear-defense system, emotions, and oxidative stress. *Redox Biol*. 2020 Jul 17;37:101588.

[147] Patil Y, Sabbu K, Iyer RB, et al. Effect of Heartfulness Meditation on Oxidative Stress and Mindfulness in Healthy Participants. *Cureus*. 2024 Jun 23;16(6):e62943.

[148] Patil Y, Sabbu K, Iyer RB, et al. Effect of Heartfulness Meditation on Oxidative Stress and Mindfulness in Healthy Participants. *Cureus*. 2024 Jun 23;16(6):e62943.

[149] Mahagita C. Roles of meditation on alleviation of oxidative stress and improvement of antioxidant system. *J Med Assoc Thailand*. 2010 Nov;93(Suppl 6):42-54.

[150] Sinatra ST, Sinatra DS, Step SWS, et al. Grounding – The universal anti-inflammatory remedy. *Biomed J.* 2023 Feb;46(1):11-16.

[151] Oschman JL, Chevalier G, Brown R. The effects of grounding (earthing) on inflammation, the immune response, wound healing, and prevention and treatment of chronic inflammatory and autoimmune diseases. *J Inflamm Res.* 2015 Mar 24;8:83–96.

[152] Hansen MM, Jones R, Tocchini K. Shinrin-Yoku (Forest Bathing) and Nature Therapy: A State-of-the-Art Review. *Int J Environ Res Public Health.* 2017 Jul 28;14(8):851.

[153] Jamil A, Gutlapalli SD, Ali M, et al. Meditation and Its Mental and Physical Health Benefits in 2023. *Cureus.* 2023 Jun 19;15(6):e40650.

[154] Sobhani V, Mokari EM, Aghajani J, et al. Islamic praying changes stress-related hormones and genes. *J Med Life.* 2022 Apr;15(4):483–488.

[155] Caliri AW, Tommasi S, Besaratinia A. Relationships among smoking, oxidative stress, inflammation, macromolecular damage, and cancer. *Mutat Res.* 2021 Jan 11;787:108365.

[156] Wu D, Cederbaum AI. Alcohol, Oxidative Stress, and Free Radical Damage. *Alcohol Res Health.* 2003;27(4):277–284.

[157] Jewell SA, Monte DD, Gentile A, et al. Decreased hepatic glutathione in chronic alcoholic patients. *J Heptaology.* 1986;3(1):1-6.

[158] Varghese J, Dakhode S. Effects of Alcohol Consumption on Various Systems of the Human Body: A Systematic Review. *Cureus.* 2022 Oct 8;14(10):e30057.

[159] Walker J, Winhusen T, Storkson J, et al. Total antioxidant capacity is significantly lower in cocaine-dependent and methamphetamine-dependent patients relative to normal controls: results from a preliminary study. *Hum Psychopharmacol.* 2014 Nov;29(6):537–543.

[160] McDonnell-Dowling K, Kelly JP. The Role of Oxidative Stress in Methamphetamine-induced Toxicity and Sources of Variation in the Design of Animal Studies. *Curr Neuropharmacol.* 2017 Feb;15(2):300–314.

[161] Berman SM, Kuczenski R, McCracken JT, et al. Potential Adverse Effects of Amphetamine Treatment on Brain and Behavior: A Review. *Mol Psychiatry.* 2008 Aug 12;14(2):123–142.

[162] Jitca G, Osz BE, Tero-Vescan A, et al. Psychoactive Drugs—From Chemical Structure to Oxidative Stress Related to Dopaminergic Neurotransmission. A Review. *Antioxidants (Basel).* 2021 Mar 4;10(3):381.

[163] Rullo L, Caputi FF, Losapio LM, et al. Effects of Different Opioid Drugs on Oxidative Status and Proteasome Activity in SH-SY5Y Cells. *Molecules.* 2022 Nov 29;27(23):8321.

[164] Packer L, Witt EH, Tritschler HJ. Alpha-lipoic acid as a biological antioxidant. *Free Radic Biol Med.* 1995;19:227-250.

[165] Wollin SD, Jones PJ. Alpha-lipoic acid and cardiovascular disease. *J Nutr.* 2003;133:3327-3330.

[166] Superti F, Russo R. Alpha-Lipoic Acid: Biological Mechanisms and Health Benefits. *Antioxidants (Basel).* 2024 Oct 12;13(10):1228.

[167] Tinggi U. Selenium: its role as antioxidant in human health. *Environ Health Prev Med.* 2008 Feb 28;13(2):102–108.

[168] Reiter RJ, Mayo JC, Tan DX, et al. Melatonin as an antioxidant: under promises but over delivers. *J Pineal Res.* 2016 Oct;61(3):253-78.

[169] Gopi S, Balakrishan P. Evaluation and clinical comparison studies on liposomal and non-liposomal ascorbic acid (vitamin C) and their enhanced bioavailability. *J Liposomal Res.* 2021;31(4):356-64.

[170] West Bengal Chemical Industries Limited. The Bioavailability Battle: How Liposomal Vitamin C Outperforms Conventional Forms in Clinical Use. Accessed May 20, 2025 from: https://www.wbcil.com/blog/the-bioavailability-battle-how-liposomal-vitamin-c-outperforms-conventional-forms-in-clinical-use/.

[171] PlantaCORP. Our liposomal vitamin C: Research and findings. Accessed May 20, 2025 from: https://plantacorp.com/our-liposomal-vitamin-c-research-and-findings/.

[172] Zmuda P, Khaidakov B, Krasowska M, et al. Bioavailability of Liposomal Vitamin C in Powder Form: A Randomized, Double-Blind, Cross-Over Trial. *Appl Sci.* 2024, 14(17), 7718

[173] Lukawski M, Dalek P, Borowik T, et al. New oral liposomal vitamin C formulation: properties and bioavailability. *J Liposome Res.* 2019 Jul;30(3):1-8.

[174] Pancorbo D, Vazquez C, Fletcher MA. Vitamin C-lipid metabolites: uptake and retention and effect on plasma C-reactive protein and oxidized LDL levels in healthy volunteers. *Med Sci Monit.* 2008 Nov;14(11):CR547-51.

[175] Righi NC, Schuch F, De Nardi AT, et al. Effects of vitamin C on oxidative stress, inflammation, muscle soreness, and strength following acute exercise: meta-analyses of randomized clinical trials. *Eur J Nutr.* 2020 Oct;59(7):2827-2839.

[176] Pancorbo D, Vazquez C, Fletcher MA. Vitamin C-lipid metabolites: uptake and retention and effect on plasma C-reactive protein and oxidized LDL levels in healthy volunteers. *Med Sci Monit.* 2008 Nov;14(11):CR547-51.

[177] Montoya-Estrada A, Garcia-Cortes AY, Romo-Yanez J, et al. The Administration of Resveratrol and Vitamin C Reduces Oxidative Stress in Postmenopausal Women-A Pilot Randomized Clinical Trial. *Nutrients.* 2024 Nov 3;16(21):3775.

[178] Zhang Q, Guo M, Chen T, et al. Walking and taking vitamin C alleviates oxidative stress and inflammation in overweight students, even in the short-term. *Front Public Health.* 2022 Oct 5:10:1024864.

[179] Uchida T, Nomura S, Oda H, Ikeda S. γ-Tocopherol Is Metabolized Faster than α-Tocopherol in Young Japanese Women. *J. Nutr. Sci. Vitaminol.* 2018;64:399–403.

[180] Szewczyk K, Chojnacka A, Gornicka M. Tocopherols and Tocotrienols-Bioactive Dietary Compounds; What Is Certain, What Is Doubt? *Int J Mol Sci.* 2021 Jun 9;22(12):6222.

[181] Yap SP, Yuen KH. Influence of lipolysis and droplet size on tocotrienol absorption from self-emulsifying formulations. *Int J Pharm.* 2004 Aug 20;281(1-2):67-78.

[182] Sepidarkish M, Akbari-Fakhrabadi M, Daneshzad E, et al. Effect of omega-3 fatty acid plus vitamin E Co-Supplementation on oxidative stress parameters: A systematic review and meta-analysis. *Clin Nutr.* 2020 Apr;39(4):1019-1025.

[183] Kim M, Eo H, Lim JG, et al. Can Low-Dose of Dietary Vitamin E Supplementation Reduce Exercise-Induced Muscle Damage and Oxidative Stress? A Meta-Analysis of Randomized Controlled Trials. *Nutrients.* 2022 Apr 12;14(8):1599.

[184] de Lima KS, Schuch FB, Righi NC, et al. Effects of the combination of vitamins C and E supplementation on oxidative stress, inflammation, muscle soreness, and muscle strength following acute physical exercise: meta-analyses of randomized controlled trials. *Crit Rev Food Sci Nutr.* 2023;63(25):7584-7597.

[185] Nguyen TTU, Yeom JH, Kim W. Beneficial Effects of Vitamin E Supplementation on Endothelial Dysfunction, Inflammation, and Oxidative Stress Biomarkers in Patients Receiving Hemodialysis: A Systematic Review and Meta-Analysis of Randomized Controlled Trials. *Int J Mol Sci.* 2021 Nov 3;22(21):11923.

[186] Salehi B, Yilmaz YB, Antika G, et al. Insights on the Use of α-Lipoic Acid for Therapeutic Purposes. *Biomolecules.* 2019 Aug 9;9(8):356.

[187] Singh U, Jialal I. Alpha-lipoic acid supplementation and diabetes. *Nutr Rev.* 2008;66(11):646-657.

[188] Hermann R, Mungo J, Cnota PJ, et al. Enantiomer-selective pharmacokinetics, oral bioavailability, and sex effects of various alpha-lipoic acid dosage forms. *Clin Pharmacol.* 2014 Nov 28:6:195-204.

[189] Geronova Research. Support Healthy Aging with Bio-Enhanced® R-Lipoic Acid. Accessed April 10, 2025 from https://geronova.com/.

[190] Brufani M. Acido α-lipoico farmaco o integratore. Una panoramica sulla farmacocinetica, le formulazioni disponibili e le evidenze cliniche nelle complicanze del diabete. *Prog Nutr.* 2014;16:62–74.

[191] Maglione E, Marrese C, Migliaro E, et al. Increasing bioavailability of (R)-alpha-lipoic acid to boost antioxidant activity in the treatment of neuropathic pain. *Acta Bio-Medica Atenei Parm.* 2015;86:226–233.

[192] Jibril AT, Jayedi A, Shab-Bidar S. Efficacy and safety of oral alpha-lipoic acid supplementation for type 2 diabetes management: a systematic review and dose–response meta-analysis of randomized trials. *Endocr Connect.* 2022 Sep 26;11(10):e220322.

[193] Rahimlou M, Asadi M, Jahromi NB, et al. Alpha-lipoic acid (ALA) supplementation effect on glycemic and inflammatory biomarkers: A Systematic Review and meta- analysis. *Clin Nutr ESPEN.* 2019 Aug;32:16-28.

[194] Konrad D, Somwar R, Sweeney G, et al. The antihyperglycemic drug alpha-lipoic acid stimulates glucose uptake via both GLUT4 translocation and GLUT4 activation: potential role of p38 mitogen-activated protein kinase in GLUT4 activation. *Diabetes.* 2001 Jun;50(6):1464-71.

[195] Vajdi M, Noshadi N, Hassanizadeh S, et al. The effects of alpha lipoic acid (ALA) supplementation on blood pressure in adults: a GRADE-assessed systematic review and dose-response meta-analysis of randomized controlled trials. *Front Cardiovasc Med.* 2023 Oct 24:10:1272837.

[196] Haghighatdoost F, Hariri M. The effect of alpha-lipoic acid on inflammatory mediators: a systematic review and meta-analysis on randomized clinical trials. *Eur J Pharmacol.* 2019 Apr 15:849:115-123.

[197] Rahimlou M, Asadi M, Jahromi NB, et al. Alpha-lipoic acid (ALA) supplementation effect on glycemic and inflammatory biomarkers: A Systematic Review and meta- analysis. *Clin Nutr ESPEN.* 2019 Aug;32:16-28.

[198] Shariffi-Zahabi E, Abdollahzad H. Alpha Lipoic Acid Supplementation and Iron Homeostasis: A Comprehensive Systematic Review and Meta-Analysis of Randomized Controlled Clinical Trials. *Int J Vitam Nutr Res.* 2024 Dec 27;95(1):36623.

[199] Zhou M, Wang L, Zhu Y. Alpha-lipoic acid alleviates oxidative stress and brain damage in patients with sevoflurane anesthesia. *Front Pharmacol.* 2025;16:1572156.

[200] Shay KP, Moreau RF, Smith EJ, et al. Alpha-lipoic acid as a dietary supplement: Molecular mechanisms and therapeutic potential. *Biochim Biophys Acta.* 2009;1790:1149–1160.

[201] Ghibu S, Richard C, Vergely C, et al. Antioxidant properties of an endogenous thiol: Alpha-lipoic acid, useful in the prevention of cardiovascular diseases. *J Cardiovasc Pharmacol.* 2009;54:391–398.

[202] Superti F, Russo R. Alpha-Lipoic Acid: Biological Mechanisms and Health Benefits. *Antioxidants (Basel).* 2024 Oct 12;13(10):1228.

[203] Esposito C, Garzarella EU, Santarcangelo C, et al. Safety and efficacy of alpha-lipoic acid oral supplementation in the reduction of pain with unknown etiology: A monocentric, randomized, double-blind, placebo-controlled clinical trial. *Biomed Pharmacol.* 2021;144:112308.

[204] Werida RH, Elshafiey RA, Ghoneim A, et al. Role of alpha-lipoic acid in counteracting paclitaxel- and doxorubicin-induced toxicities: A randomized controlled trial in breast cancer patients. *Support Care Cancer.* 2022;30:7281–7292.

[205] Borcea V, Nourooz-Zadeh J, Wolff SP, et al. alpha-Lipoic acid decreases oxidative stress even in diabetic patients with poor glycemic control and albuminuria. *Free Radic Biol Med.* 1999 Jun;26(11-12):1495-500.

[206] Mendoza-Nunez VM, Garcia-Martinez BI, Rosado-Perez J, et al. The Effect of 600 mg Alpha-lipoic Acid Supplementation on Oxidative Stress, Inflammation, and RAGE in Older Adults with Type 2 Diabetes Mellitus. *Oxid Med Cell Longev.* 2019 Jun 12:2019:3276958.

[207] Zonooz AR, Hasani M, Morvaridzadeh M, et al. Effect of alpha-lipoic acid on oxidative stress parameters: A systematic review and meta-analysis. *J Functional Foods.* 2021 Dec;87:104774.

[208] Pravst I, Aguilera JCR, Rodriguez ABC, et al. Comparative Bioavailability of Different Coenzyme Q10 Formulations in Healthy Elderly Individuals. *Nutrients*. 2020 Mar 16;12(3):784.

[209] Zmitek J, Zmitek K, Pravst I. Improving the bioavailability of coenzyme Q10 - From theory to practice. *Agro Food Ind Hi Tech*. 2008 Jul;19(4):8-10.

[210] Hidalgo-Gutierrez A, Gonzalez-Garcia P, Diaz-Casado ME, et al. Metabolic Targets of Coenzyme Q10 in Mitochondria. *Antioxidants (Basel)*. 2021 Mar 26;10(4):520.

[211] Bagheri S, Haddadi R, Saki S, et al. Neuroprotective effects of coenzyme Q10 on neurological diseases: a review article. *Front Neurosci*. 2023 Jun 23;17:1188839.

[212] Tabrizi R, Akbari M, Sharifi N, et al. The Effects of Coenzyme Q10 Supplementation on Blood Pressures Among Patients with Metabolic Diseases: A Systematic Review and Meta-analysis of Randomized Controlled Trials. *High Blood Press Cardiovasc Prev*. 2018 Mar;25(1):41-50.

[213] Zhai J, Bo Y, Lu Y, et al. Effects of Coenzyme Q10 on Markers of Inflammation: A Systematic Review and Meta-Analysis. *PLoS One*. 2017 Jan 26;12(1):e0170172.

[214] Dai S, Tian Z, Zhao D, et al. Effects of Coenzyme Q10 Supplementation on Biomarkers of Oxidative Stress in Adults: A GRADE-Assessed Systematic Review and Updated Meta-Analysis of Randomized Controlled Trials. *Antioxidants (Basel)*. 2022 Jul 13;11(7):1360.

[215] Sangsefidi ZS, Yaghoubi F, Hajiahmadi S, et al. The effect of coenzyme Q10 supplementation on oxidative stress: A systematic review and meta-analysis of randomized controlled clinical trials. *Food Sci Nutr*. 2020 Mar 19;8(4):1766-1776.

[216] Akbari A, Mobini GR, Agah S, et al. Coenzyme Q10 supplementation and oxidative stress parameters: a systematic review and meta-analysis of clinical trials. *Eur J Clin Pharmacol*. 2020 Nov;76(11):1483-1499.

[217] Gutierrex-Mariscal FM, Arenas-de Larriva AP, Limia-Perez L, et al. Coenzyme Q10 Supplementation for the Reduction of Oxidative Stress: Clinical Implications in the Treatment of Chronic Diseases. *Int J Mol Sci*. 2020 Oct 23;21(21):7870.

[218] Yin N, Harris PWR, Liu M, et al. Enhancing the Oral Bioavailability of Glutathione Using Innovative Analogue Approaches. *Pharmaceutics*. 2025 Mar 18;17(3):385.

[219] Fanelli S, Francioso A, Cavallaro RA, et al. Oral Administration of S-acetyl-glutathione: Impact on the Levels of Glutathione in Plasma and in Erythrocytes of Healthy Volunteers. *Int J Clin Nutr Bietetics*. 2018;4:IJCND-134.

[220] Fanelli S, Francioso A, Cavallaro RA, et al. Oral Administration of S-acetyl-glutathione: Impact on the Levels of Glutathione in Plasma and in Erythrocytes of Healthy Volunteers. *Int J Clin Nutr Bietetics*. 2018;4:IJCND-134.

[221] Sinha R, Sinha I, Calcagnotto A, et al. Oral supplementation with liposomal glutathione elevates body stores of glutathione and markers of immune function. *Eur J Clin Nutr*. 2018 Jan;72(1):105-111.

[222] Banjerjee PG, Paul A, Chakraborty A, et al. Liposomal glutathione: A breakthrough in cellular health by West Bengal chemical industries Ltd., Kolkata, India. *The Pharma Innovation J*. 2025;14(2):73-81.

[223] Schmidt B, Vicenzi M, Garrel C, et al. Effects of N-acetylcysteine, oral glutathione (GSH) and a novel sublingual form of GSH on oxidative stress markers: A comparative crossover study. *Redox Biol*. 2015 Jul 29;6:198–205.

[224] Richie Jr JP, Nichenametla S, Neidif W, et al. Randomized controlled trial of oral glutathione supplementation on body stores of glutathione. *Eur J Nutr*. 2015 Mar;54(2):251-63.

[225] Sinha R, Sinha I, Calcagnotto A, et al. Oral supplementation with liposomal glutathione elevates body stores of glutathione and markers of immune function. *Eur J Clin Nutr*. 2018 Jan;72(1):105-111.

[226] Dawi J, Misakyan Y, Affa S, et al. Oxidative Stress, Glutathione Insufficiency, and Inflammatory Pathways in Type 2 Diabetes Mellitus: Implications for Therapeutic Interventions. *Biomedicines*. 2024 Dec 26;13(1):18.

[227] Kalamkar S, Acharya J, Madathil AK, et al. Randomized Clinical Trial of How Long-Term Glutathione Supplementation Offers Protection from Oxidative Damage and Improves HbA1c in Elderly Type 2 Diabetic Patients. *Antioxidants (Basel).* 2022 May 23;11(5):1026.

[228] Logan AC, Wong C. Chronic fatigue syndrome: oxidative stress and dietary modifications. *Altern Med Rev.* 2001 Oct;6(5):450-9.

[229] Allen J, Bradley RD. Effects of oral glutathione supplementation on systemic oxidative stress biomarkers in human volunteers. *J Altern Complement Med.* 2011 Sep;17(9):827-33.

[230] Lana JV, Rios A, Takeyama R, et al. Nebulized Glutathione as a Key Antioxidant for the Treatment of Oxidative Stress in Neurodegenerative Conditions. *Nutrients.* 2024 Jul 31;16(15):2476.

[231] Dyck GJB, Raj P, Zieroth S, et al. The Effects of Resveratrol in Patients with Cardiovascular Disease and Heart Failure: A Narrative Review. *Int J Mol Sci.* 2019 Feb 19;20(4):904.

[232] Hou CY, Tain YL, Yu HR, et al. The Effects of Resveratrol in the Treatment of Metabolic Syndrome. *Int J Mol Sci.* 2019 Jan 28;20(3):535.

[233] Borra MT, Smith BC, Denu JM. Mechanism of human SIRT1 activation by resveratrol. *J Biol Chem.* 2005 Apr 29;280(17):17187-95.

[234] Walle T. Bioavailability of resveratrol. *Ann N Y Acad Sci.* 2011 Jan:1215:9-15.

[235] Wenzel E, Somoza V. Metabolism and bioavailability of trans-resveratrol. *Mol Nutr Food Res.* 2005 May;49(5):472-81.

[236] Johnson JJ, Nihal M, Siddiqui IA, et al. Enhancing the bioavailability of resveratrol by combining it with piperine. *Mol Nutr Food Res.* 2011 Jun 29;55(8):1169–1176.

[237] Salla M, Karaki N, Kaderi BE, et al. Enhancing the Bioavailability of Resveratrol: Combine It, Derivatize It, or Encapsulate It? *Pharmaceutics.* 2024 Apr 22;16(4):569.

[238] Shahcheraghi SH, Salemi S, Small S, et al. Resveratrol regulates inflammation and improves oxidative stress via Nrf2 signaling pathway: Therapeutic and biotechnological prospects. *Phytother Res.* 2023 Apr;37(4):1590-1605

[239] Seyyedebrahimi S, Khodabandehloo H, Esfahari EN, et al. The effects of resveratrol on markers of oxidative stress in patients with type 2 diabetes: a randomized, double-blind, placebo-controlled clinical trial. *Acta Diabetol.* 2018 Apr;55(4):341-353.

[240] Zhu P, Jin Y, Sun J, et al. The efficacy of resveratrol supplementation on inflammation and oxidative stress in type-2 diabetes mellitus patients: randomized double-blind placebo meta-analysis. *Front Endocrinol (Lausanne).* 2025 Jan 13:15:1463027.

[241] Samsamikor M, Daryani NE, Asl PR, et al. Resveratrol Supplementation and Oxidative/Anti-Oxidative Status in Patients with Ulcerative Colitis: A Randomized, Double-Blind, Placebo-controlled Pilot Study. *Arch Med Res.* 2016 May;47(4):304-9.

[242] McCormack D, McFadden D. A Review of Pterostilbene Antioxidant Activity and Disease Modification. *Oxid Med Cell Longev.* 2013 Apr 4;2013:575482.

[243] Chen Y, Zhang H, Chen Y, et al. Resveratrol and its derivative pterostilbene ameliorate intestine injury in intrauterine growth-retarded weanling piglets by modulating redox status and gut microbiota. *J Anim Sci Biotechnol.* 2021 Jun 10;12(1):70.

[244] Koh YC, Lin SJ, Hsu KY, et al. Pterostilbene Enhances Thermogenesis and Mitochondrial Biogenesis by Activating the SIRT1/PGC-1α/SIRT3 Pathway to Prevent Western Diet-Induced Obesity. *Mol Nutr Food Res.* 2023 Sep;67(18):e2300370.

[245] Chen Y, Zhang H, Ji S, et al. Resveratrol and its derivative pterostilbene attenuate oxidative stress-induced intestinal injury by improving mitochondrial redox homeostasis and function via SIRT1 signaling. *Free Radic Biol Med.* 2021 Dec:177:1-14.

[246] Riche DM, McEwen CL, Riche KD, et al. Analysis of Safety from a Human Clinical Trial with Pterostilbene. *J Toxicol.* 2013;2013:463595.

[247] Riche DM, Riche KD, Blackshear CT, et al. Pterostilbene on Metabolic Parameters: A Randomized, Double-Blind, and Placebo-Controlled Trial. *Evid Based Complement Alternat Med.* 2014 Jun 25;2014:459165.

[248] Riche DM, Deschamp D, Griswold ME, et al. Abstract 617: Impact Of Pterostilbene On Blood Pressure and Other Metabolic Parameters In Adults. *Hypertension*. 2012;60(Suppl 1).

[249] Dellinger RW, Holmes HE, Hu-Seliger T, et al. Nicotinamide riboside and pterostilbene reduces markers of hepatic inflammation in NAFLD: A double-blind, placebo-controlled clinical trial. *Hepatology*. 2023 Sep 1;78(3):863-877.

[250] Jakubczyk K, Druzga A, Katarzyna J, et al. Antioxidant Potential of Curcumin—A Meta-Analysis of Randomized Clinical Trials. *Antioxidants (Basel)*. 2020 Nov 6;9(11):1092.

[251] Peng Y, Ao M, Dong B, et al. Anti-Inflammatory Effects of Curcumin in the Inflammatory Diseases: Status, Limitations and Countermeasures. *Drug Des Devel Ther*. 2021 Nov 2;15:4503–4525.

[252] Cox FF, Misiou A, Vierkant A, et al. Protective Effects of Curcumin in Cardiovascular Diseases—Impact on Oxidative Stress and Mitochondria. *Cells*. 2022 Jan 20;11(3):342.

[253] Darvesh AS, Carroll RT, Bishayee A, et al. Curcumin and neurodegenerative diseases: a perspective. *Expert Opin Investig Drugs*. 2012 Aug;21(8):1123-40.

[254] Askarizadeh A, Barreto GE, Henney NC, et al. Neuroprotection by curcumin: A review on brain delivery strategies. *Int J Pharm*. 2020 Jul 30;585:119476.

[255] Wang P, Su C, Feng H, et al. Curcumin regulates insulin pathways and glucose metabolism in the brains of APPswe/PS1dE9 mice. *Int J Immunopathol Pharmacol*. 2017 Mar;30(1):25-43.

[256] Kulkarni SK, Dhir A. An Overview of Curcumin in Neurological Disorders. *Indian J Pharm Sci*. 2010 Mar-Apr;72(2):149–154.

[257] Bertocini-Silva C, Vlad A, Ricciarelli R, et al. Enhancing the Bioavailability and Bioactivity of Curcumin for Disease Prevention and Treatment. *Antioxidants*. 2024;13(3): 331.

[258] Kumar D, Jacob D, Subash PS, et al. Enhanced bioavailability and relative distribution of free (unconjugated) curcuminoids following the oral administration of a food-grade formulation with fenugreek dietary fibre: A randomised double-blind crossover study. *J Funct Foods*. 2016 Apr;22:578-87.

[259] IM K, Maliakel A, GG, et al. Improved blood–brain-barrier permeability and tissue distribution following the oral administration of a food-grade formulation of curcumin with fenugreek fibre. *J Funct Foods*. 2015 Apr;14:215-25.

[260] Antony B, Merina B, Iyer VS, et al. A Pilot Cross-Over Study to Evaluate Human Oral Bioavailability of BCM-95®CG (Biocurcumax™), A Novel Bioenhanced Preparation of Curcumin. *Indian J Pharm Sci*. 2008 Jul-Aug;70(4):445–449.

[261] Toden S, Theiss AL, Wang X, et al. Essential turmeric oils enhance anti-inflammatory efficacy of curcumin in dextran sulfate sodium-induced colitis. *Sci Rep*. 2017 Apr 11;7:814.

[262] Shishu mm. Comparative bioavailability of curcumin, tumeric, and Biocurcumax in traditional vichles using non-everted rat intestinal sac model. *J Functional Foods*. 2010;2:60–65.

[263] Antony B, Benny M, Rao SB. Enhancing the Absorption of Curcuminoids. Spice India. 2005:23-6.

[264] Toden S, Goel A. The Holy Grail of Curcumin and its Efficacy in Various Diseases: Is Bioavailability Truly a Big Concern? *J Restor Med*. 2017 Dec;6(1):27–36.

[265] Zhou W, Chang Y, Xiao Q, et al. Structural optimization of naturally derived Ar-turmerone, as novel neuroinflammation suppressors effective in an Alzheimer mouse model. *Biorganic Med Chem*. 2025 Jan;117:118014.

[266] Hucklenbroich J, Klein R, Neumaier B, et al. Aromatic-turmerone induces neural stem cell proliferation in vitro and in vivo. *Stem Cell Res Ther*. 2014 Sep;5:100.

[267] Tabanelli R, Brogi S, Calderone V. Improving Curcumin Bioavailability: Current Strategies and Future Perspectives. *Pharmaceutics*. 2021 Oct 17;13(10):1715.

[268] Jakubczyk K, Druzga A, Katarzyna J, et al. Antioxidant Potential of Curcumin—A Meta-Analysis of Randomized Clinical Trials. *Antioxidants (Basel)*. 2020 Nov 6;9(11):1092.

[269] Tabrizi R, Vakili S< Akbari M, et al. The effects of curcumin-containing supplements on biomarkers of inflammation and oxidative stress: A systematic review and meta-analysis of randomized controlled trials. *Phytother Res*. 2019 Feb;33(2):253-262.

[270] Dehzad MJ, Ghalandari H, Nouri M, et al. Antioxidant and anti-inflammatory effects of curcumin/turmeric supplementation in adults: A GRADE-assessed systematic review and dose-response meta-analysis of randomized controlled trials. *Cytokine*. 2023 Apr:164:156144.

[271] Sahebkar A, Serban AC, Ursoniu S, et al. Effect of curcuminoids on oxidative stress: A systematic review and meta-analysis of randomized controlled trials. *J Functional Foods*. 2015 Oct;18(Part B):898-909.

[272] Teng YS, Wu D. Anti-Fatigue Effect of Green Tea Polyphenols (-)-Epigallocatechin-3-Gallate (EGCG). *Pharmacogn Mag*. 2017 Apr 18;13(50):326–331.

[273] Hao L, Zhang A, Lv D, et al. EGCG activates Keap1/P62/Nrf2 pathway, inhibits iron deposition and apoptosis in rats with cerebral hemorrhage. *Sci Reports*. 2024 Dec 28;14:31474.

[274] Sun W, Liu X, Zhang H, et al. Epigallocatechin gallate upregulates NRF2 to prevent diabetic nephropathy via disabling KEAP1. *Free Radic Biol Med*. 2017 Jul:108:840-857.

[275] Velayutham P, Babu A, Liu D. Green Tea Catechins and Cardiovascular Health: An Update. *Curr Med Chem*. 2008;15(18):1840–1850.

[276] Widlansky ME, Hamburg NM, Anter E, et al. Acute EGCG Supplementation Reverses Endothelial Dysfunction in Patients with Coronary Artery Disease. *J Am Coll Nutr*. 2007 Apr;26(2):95–102.

[277] Chantree S, Sitticharoon C, Maikaew P, et al. Epigallocatechin gallate decreases plasma triglyceride, blood pressure, and serum kisspeptin in obese human subjects. *Exp Biol Med (Maywood)*. 2020 Oct 12;246(2):163–176.

[278] Reddy AT, Lakshmi SP, Prasad EM, et al. Epigallocatechin gallate suppresses inflammation in human coronary artery endothelial cells by inhibiting NF-κB. *Life Sci*. 2020 Oct 1:258:118136.

[279] Kapoor MP, Sugita M, Fukuzawa Y, et al. Physiological effects of epigallocatechin-3-gallate (EGCG) on energy expenditure for prospective fat oxidation in humans: A systematic review and meta-analysis. *J Nutr Biochem*. 2017 May;43:1-10.

[280] Min KJ, Kwon TK. Anticancer effects and molecular mechanisms of epigallocatechin-3-gallate. *Integr Med Res*. 2013 Dec 9;3(1):16–24.

[281] Payne A, Nahashon S< Taka E, et al. Epigallocatechin-3-Gallate (EGCG): New Therapeutic Perspectives for Neuroprotection, Aging, and Neuroinflammation for the Modern Age. *Biomolecules*. 2022 Feb 25;12(3):371.

[282] Cai S, Xie LW, Xu JY, et al. (-)-Epigallocatechin-3-Gallate (EGCG) Modulates the Composition of the Gut Microbiota to Protect Against Radiation-Induced Intestinal Injury in Mice. *Front Oncol*. 2022 Apr 11;12:848107.

[283] Tang G, Xu Y, Xhang C, et al. Green Tea and Epigallocatechin Gallate (EGCG) for the Management of Nonalcoholic Fatty Liver Diseases (NAFLD): Insights into the Role of Oxidative Stress and Antioxidant Mechanism. *Antioxidants (Basel)*. 2021 Jul 5;10(7):1076.

[284] James A, Wang K, Wang Y. Epigallocatechin Gallate (EGCG) for the Management of Nonalcoholic Fatty Liver Diseases. *Nutrients*. 2023 Jul 3;15(13):3022.

[285] Abunofal O, Mohan C. Salubrious Effects of Green Tea Catechins on Fatty Liver Disease: A Systematic Review. *Medicines (Basel)*. 2022 Mar 1;9(3):2

[286] Lup ZL, Sun HY, Wu XB, et al. Epigallocatechin-3-gallate attenuates acute pancreatitis induced lung injury by targeting mitochondrial reactive oxygen species triggered NLRP3 inflammasome activation. *Food Funct*. 2021 Jun 21;12(12):5658-5667.

[287] Adamcakova J, Mokry J. Green Tea Polyphenol (-)-Epigallocatechin-3-Gallate (EGCG): A Time for a New Player in the Treatment of Respiratory Diseases? *Antioxidants*. 2022 Aug;11(8):1566.

[288] Roychoudhury S, Agarwal A, Virk G, et al. Potential role of green tea catechins in the management of oxidative stress-associated infertility. *Reprod Biomed Online*. 2017 May;34(5):487-498.

[289] Settakorn K, Hantrakool S< Petiwathayakorn T, et al. A randomized placebo–controlled clinical trial of oral green tea epigallocatechin 3–gallate on erythropoiesis and oxidative stress in transfusion–dependent β–thalassemia patients. *Front Mol Biosci*. 2024 Jan 24;10:1248742.

[290] Shariare MH, Afnan K, Iqbal F, et al. Development and Optimization of Epigallocatechin-3-Gallate (EGCG) Nano Phytosome Using Design of Experiment (DoE) and Their In Vivo Anti-Inflammatory Studies. *Molecules*. 2020 Nov 20;25(22):5453.

[291] Gilardini L, Pasqualinotto L, Di Pierro F, et al. Effects of Greenselect Phytosome® on weight maintenance after weight loss in obese women: a randomized placebo-controlled study. *BMC Complement Altern Med*. 2016 Jul 22;16:233.

[292] Laudadio E, Minnelli C, Amici A, et al. Liposomal Formulations for an Efficient Encapsulation of Epigallocatechin-3-Gallate: An In-Silico/Experimental Approach. *Molecules*. 2018 Feb 16;23(2):441.

[293] Samavat H, Acosta LG, Wang R. Hepatotoxicity with High-Dose Green Tea Extract: Effect of Catechol-O-Methyltransferase and Uridine 5′-Diphospho-glucuronosyltransferase 1A4 Genotypes. *J Dietary Supplements*. 2022;19(5):505–520.

[294] Acosta L, Byham-Gray L, Kruzer M, et al. Hepatotoxicity with High-Dose Green Tea Extract: Effect of Catechol-O-Methyltransferase and Uridine 5'-Diphospho-glucuronosyltransferase 1A4 Genotypes. *J Diet Suppl*. 2023;20(6):850-869.

[295] Line J, Ali SE, Grice S, et al. Investigating the Immune Basis of Green Tea Extract Induced Liver Injury in Healthy Donors Expressing HLA-B35:01. *Chem Res Toxicol*. 2023 Dec 18;36(12):1872-1875.

[296] Shi Z, Zhu JX, Guo YM, et al. Epigallocatechin Gallate During Dietary Restriction — Potential Mechanisms of Enhanced Liver Injury. *Front Pharmacol*. 2021 Jan 29:11:609378.

[297] Fairweather-Tait SJ, Collings R, Hurst R. Selenium bioavailability: current knowledge and future research requirements. *Am J Clin Nutr*. 2010 May;91(5):1484S-1491S.

[298] Kuria A, Fang X, Li M, et al. Does dietary intake of selenium protect against cancer? A systematic review and meta-analysis of population-based prospective studies. *Crit Rev Food Sci Nutr*. 2020;60(4):684-694.

[299] Liang D, Liu C, Zhang X. Association between dietary selenium intake and the risk of cardiovascular disease in US adults: a population-based study. *Sci Reports*. 2025;15:13427.

[300] Hoffmann PR, Berry MJ. The influence of selenium on immune responses. *Mol Nutr Food Res*. 2008 Nov;52(11):1273–1280.

[301] Huang G, Liu Y, Zhu X, et al. Exploring the Neuroprotective Role of Selenium: Implications and Perspectives for Central Nervous System Disorders. *Exploration*. 2025 Apr 1. Online ahead of print.

[302] Ventura M, Melo M, Carrilho F. Selenium and Thyroid Disease: From Pathophysiology to Treatment. *Int J Endocrinol*. 2017 Jan 31;2017:1297658.

[303] Al-Bassman L, Shearman GC, Brocchini S, et al. The Potential of Selenium-Based Therapies for Ocular Oxidative Stress. *Pharmaceutics*. 2024 May 8;16(5):631.

[304] Sun W< Zhu J, Li S, et al. Selenium supplementation protects against oxidative stress-induced cardiomyocyte cell cycle arrest through activation of PI3K/AKT. *Metallomics*. 2020 Dec 23;12(12):1965-1978.

[305] Balali A< Sadeghi O, Khorvash F, et al. The effect of selenium supplementation on oxidative stress, clinical and physiological symptoms in patients with migraine: a double-blinded randomized clinical trial. *Front Nutr*. 2024 May 2:11:1369373.

[306] Danciu A, Ghitea TC, Bungau AF, et al. The Relationship Between Oxidative Stress, Selenium, and Cumulative Risk in Metabolic Syndrome. *In Vivo*. 2023 Nov 3;37(6):2877–2887.

[307] Richie Jr JP, Das A, Calcagnotto AM, et al. Comparative effects of two different forms of selenium on oxidative stress biomarkers in healthy men: a randomized clinical trial. *Cancer Prev Res (Phila)*. 2014 Jun 17;7(8):796–804.

[308] Zakeri N, Kelishadi MR, Asbaghi O, et al. Selenium Supplementation and Oxidative Stress: A review. *PharmaNutrition*. 2021 Sep;17:100263.

[309] Dabbagh-Bazarbachi H, Clergeaud G, Quesada IM, et al. Zinc ionophore activity of quercetin and epigallocatechin-gallate: from Hepa 1-6 cells to a liposome model. *J Agric Food Chem*. 2014 Aug 13;62(32):8085-93.

[310] Birdsall TC. Zinc picolinate: Absorption and supplementation. *Alt Med Rev*. 1996;1(1):26-30.

[311] Wegmuller R, Tay F, Aeder C, et al. Zinc absorption by young adults from supplemental zinc citrate is comparable with that from zinc gluconate and higher than from zinc oxide. *J Nutr*. 2014 Feb;144(2):132-6.

[312] Devarshi PP, Mao Q, Grant RW, et al. Comparative Absorption and Bioavailability of Various Chemical Forms of Zinc in Humans: A Narrative Review. *Nutrients*. 2024 Dec 11;16(24):4269.

[313] Wegmuller R, Tay F, Aeder C, et al. Zinc absorption by young adults from supplemental zinc citrate is comparable with that from zinc gluconate and higher than from zinc oxide. *J Nutr*. 2014 Feb;144(2):132-6.

[314] Gandia P, Bour D, Maurette JM, et al. A bioavailability study comparing two oral formulations containing zinc (Zn bis-glycinate vs. Zn gluconate) after a single administration to twelve healthy female volunteers. *Int J Vitam Nutr Res*. 2007 Jul;77(4):243-8.

[315] DiSilvestro RA, Koch E, Rakes L. Moderately High Dose Zinc Gluconate or Zinc Glycinate: Effects on Plasma Zinc and Erythrocyte Superoxide Dismutase Activities in Young Adult Women. *Biol Trace Elem Res*. 2015;168:11–14.

[316] Osko J, Pierlejewska W, Grembecka. Comparison of the Potential Relative Bioaccessibility of Zinc Supplements—In Vitro Studies. *Nutrients*. 2023 Jun 20;15(12):2813.

[317] Maares M, Haase H, et al. A Guide to Human Zinc Absorption: General Overview and Recent Advances of In Vitro Intestinal Models. *Nutrients*. 2020 Mar 13;12(3):762.

[318] Mousavi SM, Hajishafiee M, Clark CCT, et al. Clinical effectiveness of zinc supplementation on the biomarkers of oxidative stress: A systematic review and meta-analysis of randomized controlled trials. *Pharmacol Res*. 2020 Nov:161:105166.

[319] Mohammadi H, Talebi S, Ghavami A, et al. Effects of zinc supplementation on inflammatory biomarkers and oxidative stress in adults: A systematic review and meta-analysis of randomized controlled trials. *J Trace Elem Med Biol*. 2021 Dec:68:126857.

[320] Faghfouri AH, Zarezadeh M, Aghapour B, et al. Clinical efficacy of zinc supplementation in improving antioxidant defense system: A comprehensive systematic review and time-response meta-analysis of controlled clinical trials. *Eur J Pharmacol*. 2021 Sep 15:907:174243.

[321] Zarezadeh M, Faghfouri AH, Aghapour B, et al. Investigation of the clinical efficacy of Zn supplementation in improvement of oxidative stress parameters: A systematic review and dose-response meta-analysis of controlled clinical trials. *Int J Clin Practice*. 2021 Dec;75(12):e14777.

[322] Aghababaei F, Hadidi M. Recent Advances in Potential Health Benefits of Quercetin. *Pharmaceuticals (Basel)*. 2023 Jul 18;16(7):1020.

[323] Fang Y, Jin W, Guo Z, et al. Quercetin Alleviates Asthma-Induced Airway Inflammation and Remodeling through Downregulating Periostin via Blocking TGF-β1/Smad Pathway. *Pharmacology*. 2023;108(5):432-443.

[324] Townsend EA, Emala Sr. CT. Quercetin acutely relaxes airway smooth muscle and potentiates β-agonist-induced relaxation via dual phosphodiesterase inhibition of PLCβ and PDE4. *Am J Physiol Lung Cell Mol Physiol*. 2013 Jul 19;305(5):L396–L403.

[325] Mlcek J, Jurikova T, Skrovankova S, et al. Quercetin and Its Anti-Allergic Immune Response. *Molecules*. 2016 May 12;21(5):623.

[326] Cai X, Fang Z, Dou J, et al. Bioavailability of quercetin: problems and promises. *Curr Med Chem*. 2013;20(20):2572-82.

[327] Frent OD, Stefan L, Morgovan CM, et al. A Systematic Review: Quercetin—Secondary Metabolite of the Flavonol Class, with Multiple Health Benefits and Low Bioavailability. *Int J Mol Sci*. 2024;25(22):12091.

[328] Solnier J, Chang C, Roh K, et al. Quercetin LipoMicel—A Novel Delivery System to Enhance Bioavailability of Quercetin. *J Nat Health Prod Res*. 2021 Jul-Dec;3(2):1-8.

[329] Güleç K, Demirel M. Characterization and Antioxidant Activity of Quercetin/Methyl-β-Cyclodextrin Complexes. *Curr Drug Deliv*. 2016;13:444–451.

[330] Riva A, Ronchi M, Petrangolini G, et al. Improved Oral Absorption of Quercetin from Quercetin Phytosome®, a New Delivery System Based on Food Grade Lecithin. *Eur J Drug Metab Pharmacokinet*. 2018 Oct 16;44(2):169–177.

[331] Joseph A, Shanmughan P, Balakrishanan A, et al. Enhanced Bioavailability and Pharmacokinetics of a Natural Self-Emulsifying Reversible Hybrid-Hydrogel System of Quercetin: A Randomized Double-Blinded Comparative Crossover Study. *ACS Omega*. 2022 Dec;7(50):

[332] Serban MC, Sahebkar A, Zanchetti A, et al. Effects of Quercetin on Blood Pressure: A Systematic Review and Meta-Analysis of Randomized Controlled Trials. *J Am Heart Assoc*. 2016 Jul 12;5(7):e002713.

[333] Huang H, Liao D, Dong Y, et al. Effect of quercetin supplementation on plasma lipid profiles, blood pressure, and glucose levels: A systematic review and meta-analysis. *Nutr Rev*. 2020;78:615–626.

[334] Rojano-Ortega D, Pena-Amaro J, Berral-Aguilar AJ, et al. Quercetin supplementation promotes recovery after exercise-induced muscle damage: a systematic review and meta-analysis of randomized controlled trials. *Biol Sport*. 2023 Jul;40(3):813-825.

[335] Tsao JP, Bernard JR, Hsu HC, et al. Short-Term Oral Quercetin Supplementation Improves Post-exercise Insulin Sensitivity, Antioxidant Capacity and Enhances Subsequent Cycling Time to Exhaustion in Healthy Adults: A Pilot Study. *Front Nutr*. 2022 Apr;9:2022.

[336] Boots AW, Drent M, de Boer VCJ, et al. Quercetin reduces markers of oxidative stress and inflammation in sarcoidosis. *Clin Nutr*. 2011 Aug;30(4):506-12.

[337] Javadi F, Ahmadzedeh A, Eghtesadi S, et al. The Effect of Quercetin on Inflammatory Factors and Clinical Symptoms in Women with Rheumatoid Arthritis: A Double-Blind, Randomized Controlled Trial. *J Am Coll Nutr*. 2017 Jan;36(1):9-15.

[338] Chen YQ, Chen HY, Tang QQ, et al. Protective effect of quercetin on kidney diseases: From chemistry to herbal medicines. *Front Pharmacol*. 2022 Sep 2;13:968226.

[339] Frent OD, Stefan L, Morgovan CM, et al. A Systematic Review: Quercetin—Secondary Metabolite of the Flavonol Class, with Multiple Health Benefits and Low Bioavailability. *Int J Mol Sci*. 2024;25(22):12091.

[340] Alharbi OA, Alshebremi M, Babiker AY, et al. The Role of Quercetin, a Flavonoid in the Management of Pathogenesis Through Regulation of Oxidative Stress, Inflammation, and Biological Activities. *Biomolecules*. 2025;15(1):151.

[341] Hussain SA, Ahmed ZA, Mahwi TO, et al. Effect of quercetin on postprandial glucose excursion after mono- and disaccharides challenge in normal and diabetic rats. *J Diabetes Mellit*. 2012;2:82-87.

[342] Karimi A, Naeini F, Azar VA, et al. A comprehensive systematic review of the therapeutic effects and mechanisms of action of quercetin in sepsis. *Phytomedicine*. 2021 Jun:86:153567.

[343] Matias-Perez D, Antonio-Estrada C, Guerra-Martinez A, et al. Relationship of quercetin intake and oxidative stress in persistent COVID. *Front Nutr*. 2024 Jan 7;10:2023.

[344] Morka D, Mokry J, Barosova R, et al. Advances in the Use of N-Acetylcysteine in Chronic Respiratory Diseases. *Antioxidants (Basel)*. 2023 Sep 2;12(9):1713.

[345] Schwalfenberg GK. N-Acetylcysteine: A Review of Clinical Usefulness (an Old Drug with New Tricks). *J Nutr Metab*. 2021 Jun 9;2021:9949453.

[346] Tosi GM, Giustarini D, Franci L, et al. Superior Properties of N-Acetylcysteine Ethyl Ester over N-Acetyl Cysteine to Prevent Retinal Pigment Epithelial Cells Oxidative Damage. *Int J Mol Sci*. 2021 Jan 9;22(2):600.

[347] Giustarini D, Galvagni F, Dalle-Donne I, et al. N-acetylcysteine ethyl ester as GSH enhancer in human primary endothelial cells: A comparative study with other drugs. *Free Radic Biol Med*. 2018 Oct:126:202-209.

[348] Giustarini D, Milzani A, Dalle-Donne I, et al. N-Acetylcysteine ethyl ester (NACET): a novel lipophilic cell-permeable cysteine derivative with an unusual pharmacokinetic feature and remarkable antioxidant potential. *Biochem Pharmacol*. 2012 Dec 1;84(11):1522-33.

[349] Giustarini D, Milzani A, Dalle-Donne I, et al. N-Acetylcysteine ethyl ester (NACET): A novel lipophilic cell-permeable cysteine derivative with an unusual pharmacokinetic feature and remarkable antioxidant potential. *Biochem Phamracol*. 2012;84(11):1522-33.

[350] Giustarini D, Milzani A, Dalle-Donne I, et al. N-Acetylcysteine ethyl ester (NACET): A novel lipophilic cell-permeable cysteine derivative with an unusual pharmacokinetic feature and remarkable antioxidant potential. *Biochem Pharmacol*. 2012;84:1522–1533.

[351] Giustarini D, Milzani A, Dalle-Donne I, et al. N-Acetylcysteine ethyl ester (NACET): A novel lipophilic cell-permeable cysteine derivative with an unusual pharmacokinetic feature and remarkable antioxidant potential. *Biochem Pharmacol*. 2012;84:1522–1533.

[352] Lizzo G, Migliavacca E, Lamers D, et al. A Randomized Controlled Clinical Trial in Healthy Older Adults to Determine Efficacy of Glycine and N-Acetylcysteine Supplementation on Glutathione Redox Status and Oxidative Damage. *Front Aging*. 2022 Mar 7;3:852569.

[353] Kumar P, Liu C, Hsu JW, et al. Glycine and N-acetylcysteine (GlyNAC) supplementation in older adults improves glutathione deficiency, oxidative stress, mitochondrial dysfunction, inflammation, insulin resistance, endothelial dysfunction, genotoxicity, muscle strength, and cognition: Results of a pilot clinical trial. *Clin Transl Med*. 2021 Mar;11(3):e372.

[354] Sekhar RV. GlyNAC Supplementation Improves Glutathione Deficiency, Oxidative Stress, Mitochondrial Dysfunction, Inflammation, Aging Hallmarks, Metabolic Defects, Muscle Strength, Cognitive Decline, and Body. *J Nutr*. 2021 Dec 3;151(12):3606-3616.

[355] Kumar P, Liu C, Suliburk J, et al. Supplementing Glycine and N-Acetylcysteine (GlyNAC) in Older Adults Improves Glutathione Deficiency, Oxidative Stress, Mitochondrial Dysfunction, Inflammation, Physical Function, and Aging Hallmarks: A Randomized Clinical Trial.

[356] Faghfouri AH, Zarezadeh M, Tavakoli-Rouzbehani OM, et al. The effects of N-acetylcysteine on inflammatory and oxidative stress biomarkers: A systematic review and meta-analysis of controlled clinical trials. *Eur J Pharmacol*. 2020 Oct 5:884:173368.

[357] Elgar K. N-acetylcysteine: A Review of Clinical Use and Efficacy. *Nutr Med J*. 2022;1(3):26-45.

[358] Cazzola M, Calzetta L, Page CP, et al. Influence of N-acetylcysteine on chronic bronchitis or COPD exacerbations: A meta-analysis. *Eur Respiratory Rev*. 2015;24(137):451-61.

[359] Huang C, Kup S, Lin L, et al. The efficacy of N-acetylcysteine in chronic obstructive pulmonary disease patients: a meta-analysis. *Ther Adv Respir Dis*. 2023 Mar 17;17:17534666231158563.

[360] Jiang C, Zou J, Lv Q, et al. Systematic review and meta-analysis of the efficacy of N-acetylcysteine in the treatment of acute exacerbation of chronic obstructive pulmonary disease. *Ann Palliat Med*. 2021 Jun;10(6):6564-76.

[361] Zheng W, Zhang QE, Cai DB, et al. N-acetylcysteine for major mental disorders: a systematic review and meta-analysis of randomized controlled trials. *Acta Psychiatr Scand*. 2018 May;137(5):391-400.

[362] Smaga I, Frankowska M, Filip M. N-acetylcysteine as a new prominent approach for treating psychiatric disorders. *British J Pharmacol*. 2021 Jul;178(13):2569-94.

[363] Ooi SL, Green R, Pak SC. N-Acetylcysteine for the Treatment of Psychiatric Disorders: A Review of Current Evidence. *Biomed Res Int*. 2018 Oct 22;2018:2469486.

[364] Dean O, Giorlando F, Berk M. N-acetylcysteine in psychiatry: current therapeutic evidence and potential mechanisms of action. *J Psych Neurosci*. 2011;36(2):78–86.

[365] Jannatifar R, Parivar K, Roodbari NH, et al. Effects of N-acetyl-cysteine supplementation on sperm quality, chromatin integrity and level of oxidative stress in infertile men. *Reprod Biol Endocrinol*. 2019 Feb 16;17:24.

[366] Thakker D, Raval A, Patel I, et al. N-Acetylcysteine for Polycystic Ovary Syndrome: A Systematic Review and Meta-Analysis of Randomized Controlled Clinical Trials. *Obstet Gynecol Int*. 2015 Jan 8;2015:817849.

[367] Aljohani W, Pak B, Chan H, et al. Role of N-Acetylcysteine in the Treatment of Acute Nonacetaminophen, Nonalcoholic and Nonviral Hepatitis: A Meta-analysis. *J Canadian Assoc Gastroenterol*. 2021 Jun;4(3):125-30.

[368] Amjad W, Thuluvath P, Mansoor M, et al. N-acetylcysteine in non-acetaminophen-induced acute liver failure: a systematic review and meta-analysis of prospective studies. *Prz Gastroenterol*. 2021 Jul 14;17(1):9–16.

[369] Nikbaf-Shandiz M, Adeli S, Faghfouri AH, et al. The efficacy of N-acetylcysteine in improving liver function: A systematic review and meta-analysis of controlled clinical trials. *PharmaNutr*, 2023 Jun;24:100343.

[370] Giannaccare G, Pellegrini M, Senni C, et al. Clinical Applications of Astaxanthin in the Treatment of Ocular Diseases: Emerging Insights. *Mar Drugs*. 2020 May 1;18(5):239.

[371] Ito N, Seki S, Ueda F. The Protective Role of Astaxanthin for UV-Induced Skin Deterioration in Healthy People—A Randomized, Double-Blind, Placebo-Controlled Trial. *Nutrients*. 2018 Jun 25;10(7):817.

[372] Choi HD, Youn KY, Shin WG. Positive effects of astaxanthin on lipid profiles and oxidative stress in overweight subjects. *Plant Foods Hum Nutr*. 2011 Nov;66(4):363-9.

[373] Chang MX, Xiong F. Astaxanthin and its Effects in Inflammatory Responses and Inflammation-Associated Diseases: Recent Advances and Future Directions. *Molecules*. 2020 Nov 16;25(22):5342.

[374] Odeberg JM, Lignell A, Pettersson A, et al. Oral bioavailability of the antioxidant astaxanthin in humans is enhanced by incorporation of lipid based formulations. *Eur J Pharm Sci*. 2003 Jul;19(4):299-304.

[375] Choi HD, Kang HE, Yang SH, et al. Pharmacokinetics and First-Pass Metabolism of Astaxanthin in Rats. *Br J Nutr*. 2011;105:220–227.

[376] Pan L, Meng H, Li J, et al. Enhancement of Astaxanthin Bioaccessibility by Encapsulation in Liposomes: An In Vitro Study. *Molecules*. 2024;29(8):1687.

[377] Chang HI, Shao CW, Huang E, et al. Development of Astaxanthin-Loaded Nanosized Liposomal Formulation to Improve Bone Health. *Pharmaceuticals (Basel)*. 2022 Apr 18;15(4):490.

[378] Wahab BRA, Affandi MMRMM, Fakurazi S, et al. Nanocarrier System: State-of-the-Art in Oral Delivery of Astaxanthin. *Antioxidants (Basel)*. 2022 Aug 27;11(9):1676.

[379] Madhavi D, Kagan D, Seshadri S. A Study on the Bioavailability of a Proprietary, Sustained-release Formulation of Astaxanthin. *Integr Med (Encinitas)*. 2018 Jun;17(3):38–42.

[380] Shen X, Fang T, Zheng J, et al. Physicochemical Properties and Cellular Uptake of Astaxanthin-Loaded Emulsions. *Molecules*. 2019 Feb 18;24(4):727.

[381] University of South Florida. Astaxanthin Formulation Bioavailability. Accessed April 8, 2025 form: https://ichgcp.net/clinical-trials-registry/NCT02397811.

[382] Landrum J, Bone R, Mendez V, et al. Comparison of dietary supplementation with lutein diacetate and lutein: a pilot study of the effects on serum and macular pigment. *Acta Biochimica Polonica*. 2012;59(1):167–9.

[383] Norkus EP, Norkus KL, Dharmarajan TS, et al. Serum lutein response is greater from free lutein than from esterified lutein during 4 weeks of supplementation in healthy adults. J Am Coll Nutr. 2010;29(6):575-85.

[384] Zhou X, Cao Q, Orfila C, et al. Systematic Review and Meta-Analysis on the Effects of Astaxanthin on Human Skin Ageing. *Nutrients*. 2021 Aug 24;13(9):2917.

[385] Ma B, Lu J, Kang T, et al. Astaxanthin supplementation mildly reduced oxidative stress and inflammation biomarkers: a systematic review and meta-analysis of randomized controlled trials. *Nutr Res*. 2022 Mar:99:40-50.

[386] Wu D, Xu H, Chen J, et al. Effects of Astaxanthin Supplementation on Oxidative Stress. *Int J Vitam Nutr Res*. 2020 Jan;90(1-2):179-194.

[387] Davinelli S, Nielsen ME, Scapagnini G. Astaxanthin in Skin Health, Repair, and Disease: A Comprehensive Review. *Nutrients*. 2018 Apr 22;10(4):522.

[388] Mokhtari E, Rafiei S, Shokri-Mashhadi N, et al. Impact of astaxanthin supplementation on blood pressure: A systematic review and meta-analysis of randomized controlled trials. *J Funct Foods*. 2021 Dec;87:104860.

[389] Pereira CPM, Souza ACR, Vasconcelos AR, et al. Antioxidant and anti-inflammatory mechanisms of action of astaxanthin in cardiovascular diseases (Review). *Int J Mol Med*. 2021 Jan;47(1):37-48.

[390] Heidari M, Chaboksafar M, Alizadeh M, et al. Effects of Astaxanthin supplementation on selected metabolic parameters, anthropometric indices, Sirtuin1 and TNF-α levels in patients with coronary artery disease: A randomized, double-blind, placebo-controlled clinical trial. *Front Nutr*. 2023 Mar 27:10:1104169.

[391] Queen CJJ, Sparks SA, Marchant DC, et al. The Effects of Astaxanthin on Cognitive Function and Neurodegeneration in Humans: A Critical Review. *Nutrients*. 2024;16(6):826.

[392] Giannaccare G, Pellegrini M, Senni C, et al. Clinical Applications of Astaxanthin in the Treatment of Ocular Diseases: Emerging Insights. *Mar Drugs*. 2020 May 1;18(5):239.

[393] Tian L, Wen Y, Li S, et al. Benefits and Safety of Astaxanthin in the Treatment of Mild-To-Moderate Dry Eye Disease. *Front Nutr*. 2022 Jan 13:8:796951.

[394] Cristaldi M, Anfuso CD, Spampinato G, et al. Comparative Efficiency of Lutein and Astaxanthin in the Protection of Human Corneal Epithelial Cells In Vitro from Blue-Violet Light Photo-Oxidative Damage. *Appl Sci*. 2022;12(3):1268.

[395] Sajovic J, Meglic A, Glavac D, et al. The Role of Vitamin A in Retinal Diseases. *Int J Mol Sci*. 2022 Jan 18;23(3):1014.

[396] VanBuren A, Everts HB. Vitamin A in Skin and Hair: An Update. *Nutrients*. 2022 Jul 19;14(14):2952.

[397] Huang Z, Liu Y, Qi G, et al. Role of Vitamin A in the Immune System. *J Clin Med*. 2018 Sep 6;7(9):258.

[398] National Institutes of Health, Office of Dietary Supplements. Vitamin A and Carotenoids. Accessed April 9, 2025 from: https://ods.od.nih.gov/factsheets/VitaminA-HealthProfessional/.

[399] National Institutes of Health, Office of Dietary Supplements. Vitamin A and Carotenoids. Accessed April 9, 2025 from: https://ods.od.nih.gov/factsheets/VitaminA-HealthProfessional/.

[400] Lee S, Huh I, Kang S, et al. Meta-Analysis of Randomized Clinical Trials Evaluating Effectiveness of a Multivitamin Supplementation against Oxidative Stress in Healthy Subjects. *Nutrients*. 2022;14(6):1170.

[401] Miazek K, Beton K, Sliwinska A, et al. The Effect of β-Carotene, Tocopherols and Ascorbic Acid as Anti-Oxidant Molecules on Human and Animal In Vitro/In Vivo Studies: A Review of Research Design and Analytical Techniques Used. *Biomolecules*. 2022 Aug 7;12(8):1087.

[402] Zhang Y, Ding J, Liang J. Associations of Dietary Vitamin A and Beta-Carotene Intake With Depression. A Meta-Analysis of Observational Studies. *Front Nutr*. 2022 Apr 25;9:881139.

[403] Ji N, Lei M, Chen Y, et al. How Oxidative Stress Induces Depression? *ASN Neuro*. 2023 Jan-Dec:15:17590914231181037.

[404] Malekahmadi M, Moghaddam OM, Firouzi S, et al. Effects of pycnogenol on cardiometabolic health: A systematic review and meta-analysis of randomized controlled trials. *Pharmacol Res*. 2019 Dec:150:104472.

[405] Nattagh-Eshtivani E, Gheflati A, Barghchi H, et al. The role of Pycnogenol in the control of inflammation and oxidative stress in chronic diseases: Molecular aspects. *Phytother Res*. 2022 Jun;36(6):2352-2374.

[406] Marini A, Grether-Beck S, Jaenicke T, et al. Pycnogenol® effects on skin elasticity and hydration coincide with increased gene expressions of collagen type I and hyaluronic acid synthase in women. *Skin Pharmacol Physiol*. 2012;25(2):86-92.

[407] Grimm T, Skrabala R, Chovanova Z, et al. Single and multiple dose pharmacokinetics of maritime pine bark extract (Pycnogenol) after oral administration to healthy volunteers. *BMC Clin Pharmacol*. 2006 Aug 3;6:4.

[408] Weichmann F, Rohdewald P. Pycnogenol® French maritime pine bark extract in randomized, double-blind, placebo-controlled human clinical studies. *Front Nutr*. 2024 May 2;11:1389374.

[409] Rhodewald P. A review of the French maritime pine bark extract (Pycnogenol), a herbal medication with a diverse clinical pharmacology. *Int J Clin Pharmacol Ther*. 2002 Apr;40(4):158-68.

[410] Weichmann F, Rohdewald P. Pycnogenol® French maritime pine bark extract in randomized, double-blind, placebo-controlled human clinical studies. *Front Nutr*. 2024 May 2;11:1389374.

[411] Schoonees A, Visser J, Musekiwa A, et al. Pycnogenol(®) for the treatment of chronic disorders. *Cochrane Database Syst Rev*. 2012 Feb 15:(2):CD008294.

[412] Mohammadi S, Fulop T, Khalil A, et al. Does supplementation with pine bark extract improve cardiometabolic risk factors? A systematic review and meta-analysis. *BMC Complement Med Ther*. 2025 Feb 22;25:71.

[413] Zhang Z, Tong X, Wei Yl, et al. Effect of Pycnogenol Supplementation on Blood Pressure: A Systematic Review and Meta-analysis. *Iran J Public Health*. 2018 Jun;47(6):779–787.

[414] Malekahmadi M, Moghaddam OM, Firouzi S, et al. Effects of pycnogenol on cardiometabolic health: A systematic review and meta-analysis of randomized controlled trials. *Pharmacol Res*. 2019 Dec:150:104472

[415] Mohammadi S, Fulop T, Khalil A, et al. Does supplementation with pine bark extract improve cardiometabolic risk factors? A systematic review and meta-analysis. *BMC Complement Med Ther*. 2025 Feb 22;25:71.

[416] Mohammadi S, Fulop T, Khalil A, et al. Does supplementation with pine bark extract improve cardiometabolic risk factors? A systematic review and meta-analysis. *BMC Complement Med Ther*. 2025 Feb 22;25:71.

[417] Weichmann F, Rohdewald P. Pycnogenol® French maritime pine bark extract in randomized, double-blind, placebo-controlled human clinical studies. *Front Nutr*. 2024 May 2;11:1389374.

[418] Simpsom T, Kure C, Stough C. Assessing the Efficacy and Mechanisms of Pycnogenol® on Cognitive Aging From In Vitro Animal and Human Studies. *Front Pharmacol*. 2019 Jul 3;10:694.

[419] Weichmann F, Rohdewald P. Pycnogenol® French maritime pine bark extract in randomized, double-blind, placebo-controlled human clinical studies. *Front Nutr*. 2024 May 2;11:1389374.

[420] Belcaro G, Luzzi R, Dugall M, et al. Pycnogenol® improves cognitive function, attention, mental performance and specific professional skills in healthy professionals aged 35-55. *J Neurosurg Sci*. 2014 Dec;58(4):239-48.

[421] Ryan J Croft K, Mori T, et al. An examination of the effects of the antioxidant Pycnogenol on cognitive performance, serum lipid profile, endocrinological and oxidative stress biomarkers in an elderly population. *J Psychopharmacol*. 2008 Jul;22(5):553-62.

[422] Belcaro G, Duggal M, Ippolito E, et al. The COFU3 Study. Improvement in cognitive function, attention, mental performance with Pycnogenol® in healthy subjects (55-70) with high oxidative stress. *J Neurosurg Sci*. 2015 Dec;59(4):437-46.

[423] Simpsom T, Kure C, Stough C. Assessing the Efficacy and Mechanisms of Pycnogenol® on Cognitive Aging From In Vitro Animal and Human Studies. *Front Pharmacol*. 2019 Jul 3;10:694.

[424] Nattagh-Eshtivani E, Gheflati A, Barghchi H, et al. The role of Pycnogenol in the control of inflammation and oxidative stress in chronic diseases: Molecular aspects. *Phytother Res*. 2022 Jun;36(6):2352-2374.

[425] Schoonees A, Visser J, Musekiwa A, et al. Pycnogenol(®) for the treatment of chronic disorders. *Cochrane Database Syst Rev*. 2012 Feb 15:(2):CD008294.

[426] Domanico D, Fragiotta S, Cutini A, et al. Circulating levels of reactive oxygen species in patients with nonproliferative diabetic retinopathy and the influence of antioxidant supplementation: 6-month follow-up. *Ind J Ophthalmol*. 2015;63:9–14.

[427] Lau BHS, Riesen SK, Truong KP, et al. Pycnogenol as an adjunct in the management of childhood asthma. *J Asthma*. 2004;41(8):825-32.

[428] Belcaro G, Luzzi R, Cesinaro Di Rocco P, et al. Pycnogenol® improvements in asthma management. *Panminerva Med*. 2011 Sep;53(3 Suppl 1):57-64.

[429] Nattagh-Eshtivani E, Gheflati A, Barghchi H, et al. The role of Pycnogenol in the control of inflammation and oxidative stress in chronic diseases: Molecular aspects. *Phytother Res*. 2022 Jun;36(6):2352-2374.

[430] Schoonees A, Visser J, Musekiwa A, et al. Pycnogenol(®) for the treatment of chronic disorders. *Cochrane Database Syst Rev*. 2012 Feb 15:(2):CD008294.

[431] Farid R, Mirfeizi Z, Mirheidai M, et al. Pycnogenol supplementation reduces pain and stiffness and improves physical function in adults with knee osteoarthritis. *Nutr Res*. 2007;27:692–697.

[432] Saller R, Brignoli R, Melzer J, et al. An updated systematic review with meta-analysis for the clinical evidence of silymarin. *Forsch Komplementmed*. 2008 Feb;15(1):9-20.

[433] Gillessen A, Schmidt HHJ. Silymarin as Supportive Treatment in Liver Diseases: A Narrative Review. *Adv Ther*. 2020 Feb 17;37(4):1279–1301.

[434] Vostlalova J, Tinkova E, Biedermann D, et al. Skin Protective Activity of Silymarin and its Flavonolignans. *Molecules*. 2019 Mar 14;24(6):1022.

[435] Boira C, Chapuis E, Scandolera A, et al. Silymarin Alleviates Oxidative Stress and Inflammation Induced by UV and Air Pollution in Human Epidermis and Activates β-Endorphin Release through Cannabinoid Receptor Type 2. *Cosmetics*. 2024;11(1):30.

[436] Javed S, Kohli K, Ali A. Reassessing Bioavailability of Silymarin. *Alternat Med Rev*. 2011 Sep;16(3):239-49.

[437] Wu JW, Lin LC, Hung SC, et al. Analysis of silibinin in rat plasma and bile for hepatobiliary excretion and oral bioavailability application. *J Pharm Biomed Anal*. 2007;45(4):635–641.

[438] Selc M, Macova R, Babeloca A. Novel Strategies Enhancing Bioavailability and Therapeutical Potential of Silibinin for Treatment of Liver Disorders. *Drug Des Devel Ther*. 2024 Oct 19;18:4629–4659.

[439] Barzaghi N, Crema F, Gatti G, et al. Pharmacokinetic studies on IdB 1016, a silybin-phosphatidylcholine complex, in healthy human subjects. *Eur J Drug Metab Pharmacokinet*. 1990 Oct-Dec;15(4):333-8.

[440] Morazzoni P, Magistretti MJ, Giachetti C, et al. Comparative bioavailability of Silipide, a new flavanolignan complex, in rats. *Eur J Drug Metab Pharmacokinet*, 1992;17:39-44.

[441] Nahum MS, Miguel DM, Jahir SN, et al. Superior silybin bioavailability of silybin–phosphatidylcholine complex in oily-medium soft-gel capsules versus conventional silymarin tablets in healthy volunteers*. *BMC Pharmacol Toxicol*. 2019 Jan 11;20:5.

[442] Kumar N, Rai A, Reddy ND, et al. Silymarin liposomes improves oral bioavailability of silybin besides targeting hepatocytes, and immune cells. *Pharmacol Rep*. 2014 Oct;66(5):788-98.

[443] Yang G, Zhao Y, Zhang Y, et al. Enhanced oral bioavailability of silymarin using liposomes containing a bile salt: Preparation by supercritical fluid technology and evaluation in vitro and in vivo. *Int J.Nanomed*. 2015;10:6633–6644.

[444] Zhu B, Ding Z, Rong X, et al. Silybin Cocrystals with Improved Solubility and Bioavailability. *Pharmaceuticals*. 2025;18(1):90.

[445] Seo SR, Kim GY, Kim MH, et al. Nanocrystal Formulation to Enhance Oral Absorption of Silybin: Preparation, In Vitro Evaluations, and Pharmacokinetic Evaluations in Rats and Healthy Human Subjects. *Pharmaceutics*. 2024;16(8):1033.

[446] Poruba M, Kazdova L, Oliyarnyk O, et al. Improvement bioavailability of silymarin ameliorates severe dyslipidemia associated with metabolic syndrome. *Xenobiotica*. 2015;45(9):751-6.

[447] Javed S, Kohli K, Ahsan W. Bioavailability augmentation of silymarin using natural bioenhancers: An in vivo pharmacokinetic study. *Braz J Pharm Sci*. 2022;58.

[448] Vajdi M, Adeli S, Karimi A, et al. The Impact of Silymarin on Inflammation and Oxidative Stress: A Systematic Review and Meta-Analysis of Randomized Controlled Trials. *Int J Clin Pract*. 2025 Jan:3985207.

[449] Bahari H, Jazinaki MS, Rashidmayan M, et al. The effects of silymarin consumption on inflammation and oxidative stress in adults: a systematic review and meta-analysis. *Inflammopharmacology*. 2024 Apr;32(2):949-963.

[450] Mohammadi S, Ashtary-Laky D, Asbaghi O, et al. Effects of silymarin supplementation on liver and kidney functions: A systematic review and dose-response meta-analysis. *Phytother Res*. 2024 May;38(5):2572-2593.

[451] Mohammadi S, Asbaghi O, Afrsham R, et al. Impacts of Supplementation with Silymarin on Cardiovascular Risk Factors: A Systematic Review and Dose–Response Meta-Analysis. *Antioxidants (Basel)*. 2024 Mar 24;13(4):390.

[452] Vanekova Z, Rollinger JM. Bilberries: Curative and Miraculous – A Review on Bioactive Constituents and Clinical Research. *Front Pharmacol*. 2022 Jun 29;13:909914.

[453] Bryl-Gorecka P, Sathanoori R, et al. Bilberry Supplementation after Myocardial Infarction Decreases Microvesicles in Blood and Affects Endothelial Vesiculation. *Mol Nutr Food Res*. 2020 Sep 10;64(20):2000108.

[454] Sharma A, Lee HJ. Anti-Inflammatory Activity of Bilberry (Vaccinium myrtillus L.). *Curr Issues Mol Biol*. 2022 Sep 30;44(10):4570–4583.

[455] Mueller D, Jung K, Winter M, et al. Encapsulation of anthocyanins from bilberries – Effects on bioavailability and intestinal accessibility in humans. *Food Chem*. 2018 May 15;248:217-24.

[456] Mueller D, Jung K, Winter M, et al. Human intervention study to investigate the intestinal accessibility and bioavailability of anthocyanins from bilberries. *Food Chem*. 2017 Sep 15;231:275-86.

[457] Mueller D, Jung K, Winter M, et al. Encapsulation of anthocyanins from bilberries – Effects on bioavailability and intestinal accessibility in humans. *Food Chem*. 2018 May 15;248:217-24.

[458] Stote KS, Burns G, Mears K, et al. The Effect of Berry Consumption on Oxidative Stress Biomarkers: A Systematic Review of Randomized Controlled Trials in Humans. *Antioxidants*. 2023;12(7):1443.

[459] Chan SW, Tomlinson B. Effects of Bilberry Supplementation on Metabolic and Cardiovascular Disease Risk. *Molecules*. 2020 Apr 3;25(7):1653.

[460] Chan SW, Chu TTW, Choi SW, et al. Impact of short-term bilberry supplementation on glycemic control, cardiovascular disease risk factors, and antioxidant status in Chinese patients with type 2 diabetes. *Phytother Res*. 2021 Jun;35(6):3236-3245.

[461] Bohn SK, Myhrstad MCW, Thoresen M, et al. Bilberry/red grape juice decreases plasma biomarkers of inflammation and tissue damage in aged men with subjective memory impairment –a randomized clinical trial. *BMC Nutr*. 2021;7(75):2021.

[462] Sharma A, Lee HJ. Anti-Inflammatory Activity of Bilberry (Vaccinium myrtillus L.). *Curr Issues Mol Biol*. 2022;44(10):4570-4583.

[463] Aboonabi A, Meyer RR, Gaiz A, et al. Anthocyanins in berries exhibited anti-atherogenicity and antiplatelet activities in a metabolic syndrome population. *Nutr Res*. 2020, 76, 82–93.

[464] Kapala A, Szlendak M, Motacka E. The Anti-Cancer Activity of Lycopene: A Systematic Review of Human and Animal Studies. *Nutrients*. 2022 Dec 3;14(23):5152.

[465] Przybylska S, Tokarcyzk G. Lycopene in the Prevention of Cardiovascular Diseases. *Int J Mol Sci*. 2022 Feb 10;23(4):1957.

[466] Boileau AC, Merchen NR, Wasson K, et al. Cis-lycopene, is more bioavailable than trans-lycopene in vitro and in vivo in lymph-cannulated ferrets. *J Nutr*. 1999;129(6):1176-81.

[467] Arballo J, Amengual J, Erdman Jr JW. Lycopene: A Critical Review of Digestion, Absorption, Metabolism, and Excretion. *Antioxidants (Basel)*. 2021 Feb 25;10(3):342.

[468] Arballo J, Amengual J, Erdman Jr JW. Lycopene: A Critical Review of Digestion, Absorption, Metabolism, and Excretion. *Antioxidants (Basel)*. 2021 Feb 25;10(3):342.

[469] Li X, Xu J. Lycopene Supplement and Blood Pressure: An Updated Meta-Analysis of Intervention Trials. *Nutrients*. 2013 Sep;5(9):3696-3712.

[470] Bin-Jumah MN, Nadeem MS, Filani SJ, et al. Lycopene: A Natural Arsenal in the War against Oxidative Stress and Cardiovascular Diseases. *Antioxidants*. 2022;11(2):232.

[471] Przybylska S, Tokarcyzk G. Lycopene in the Prevention of Cardiovascular Diseases. *Int J Mol Sci*. 2022 Feb 10;23(4):1957.

[472] Rissanen T, Voutilainen S, Nyyssonen K, et al. Lycopene, atherosclerosis, and coronary heart disease. *Exp Biol Med (Maywood)*. 2002 Nov;227(10):900-7.

[473] Rejali L, Ozumerzifon S, Nayeri H, et al. Risk reduction and prevention of cardiovascular diseases: biological mechanisms of lycopene. *Bioactive Comp Health Dis*. 2022;5(10):202-11.

[474] Bin-Jumah MN, Nadeem MS, Gilani SJ, et al. Lycopene: A Natural Arsenal in the War against Oxidative Stress and Cardiovascular Diseases. *Antioxidants*. 2022;11(2):232.

[475] Thies F, Mills LM, Moir S, et al. Cardiovascular benefits of lycopene: fantasy or reality? *Proc Nutr Soc*. 2017 May;76(2):122-129.

[476] Mozon I, Stoian D, Caraba A, et al. Lycopene and Vascular Health. *Front Pharmacol*. 2018 May 23;9:521.

[477] Zhong Q, Piao YY, Yin S, et al. Association of serum lycopene concentrations with all-cause and cardiovascular mortality among individuals with chronic kidney disease: A cohort study. *Front Nutr*. 2022 Dec 5:9:1048884.

[478] Balali A, Fathzadeh K, Askari G, et al. Dietary intake of tomato and lycopene, blood levels of lycopene, and risk of total and specific cancers in adults: a systematic review and dose–response meta-analysis of prospective cohort studies. *Front Nutr*. 2025 Feb 12:12:1516048.

[479] Arballo J, Amengual J, Erdman Jr JW. Lycopene: A Critical Review of Digestion, Absorption, Metabolism, and Excretion. Antioxidants (Basel). 2021 Feb 25;10(3):342.

[480] Imran M, Ghorat F, Ul-Haq I, et al. Lycopene as a Natural Antioxidant Used to Prevent Human Health Disorders. *Antioxidants*. 2020;9(8):706.

[481] Kim MJ, Kim H. Anticancer Effect of Lycopene in Gastric Carcinogenesis. *J Cancer Prev*. 2015;20(2):92-96.

[482] Inoue T, Yoshida K, Sasaki E, et al. Effect of Lycopene Intake on the Fasting Blood Glucose Level: A Systematic Review with Meta-Analysis. *Nutrients*. 2023;15(1):122.

[483] Kulawik A, Cielecka-Piontek J, Czerny B, et al. The Relationship Between Lycopene and Metabolic Diseases. *Nutrients*. 2024;16(21):3708.

[484] Goenawan H, Pratiwi YS, Dewi NP, et al. Beneficial Effect of Lycopene on Diabetes Mellitus and its Possible Mechanism: A Review. *Trop J Nat Prod Res*. 2021 Mar;5(3):420-33.

[485] Kulawik A, Cielecka-Piontek J, Czerny B, et al. What Do We Know about the Relationship between Lycopene and Metabolic Diseases? Review. 2024 Sep. Online ahead of print.

[486] Kulawik A, Cielecka-Piontek J, Czerny B, et al. The Relationship Between Lycopene and Metabolic Diseases. *Nutrients*. 2024;16(21):3708.

[487] Crowe-White KM, Phillips TA, Ellis AC. Lycopene and cognitive function. *J Nutr Sci*. 2019 May 29;8:e20.

[488] Lopez-Valerde N, Lopez-Valerde A, de Sousa BM, et al. Systematic review and meta-analysis of the antioxidant capacity of lycopene in the treatment of periodontal disease. *Front Bioeng Biotechnol.* 2024 Jan 8:11:1309851.

[489] Groten K, Marini A, Grether-Beck S, et al. Tomato Phytonutrients Balance UV Response: Results from a Double-Blind, Randomized, Placebo-Controlled Study. *Skin Pharmacol Physiol.* 2019;32(2):101–108.

[490] Zhang N, Kong F, Zhao L, et al. Essential oil, juice, and ethanol extract from bergamot confer improving effects against primary dysmenorrhea in rats. *J Food Biochem.* 2021 Feb;45(2):e13614.

[491] Catalano R, Procopio F, Chavarria D, et al. Molecular Modeling and Experimental Evaluation of Non-Chiral Components of Bergamot Essential Oil with Inhibitory Activity against Human Monoamine Oxidases. *Molecules.* 2022;27:2467.

[492] Tian L, Hu T, Zhang S, Zhang H, et al. A Comparative Study on Relieving Exercise-Induced Fatigue by Inhalation of Different Citrus Essential Oils. *Molecules.* 2022 May 18;27(10):3239.

[493] Lin L, Yang Q, Li TT, et al. Effect of aromatherapy in patients with Alzheimer's disease: a randomised controlled clinical trial. 2022 Apr. Online ahead of print.

[494] Dadkhah A, Fatemi F. Heart and kidney oxidative stress status in septic rats treated with caraway extracts. *Pharm Biol.* 2011 Jul;49(7):679-86.

[495] Tavares LA, Rezende AA, Santos JL, et al. Cymbopogon winterianus Essential Oil Attenuates Bleomycin-Induced Pulmonary Fibrosis in a Murine Model. *Pharmaceutics.* 2021 May 9;13(5):679.

[496] Bakour M, Soulo N, Hammas N, et al. The Antioxidant Content and Protective Effect of Argan Oil and Syzygium aromaticum Essential Oil in Hydrogen Peroxide-Induced Biochemical and Histological Changes. *Int J Mol Sci.* 2018 Feb 18;19(2).

[497] Campos C, de Castro AL, Tavares AMV, et al. Effect of Free and Nanoencapsulated Copaiba Oil on Monocrotaline-induced Pulmonary Arterial Hypertension. *J Cardiovasc Pharmacol.* 2017 feb;69(2):79-85.

[498] Jana S, Patra K, Sarkar S, et al. Antitumorigenic potential oil linalool is accompanied by modulation of oxidative stress: an in vivo study in sarcoma-180 solid tumor. *Nutr Cancer.* 2014;66(5):835-48.

[499] Morovati A, Pourghassem Gargari B, Sarbakhsh P, et al. The effect of cumin supplementation on metabolic profiles in patients with metabolic syndrome: A randomized, triple blind, placebo-controlled clinical trial. *Phytother Res.* 2019 Apr;33(4):1182-1190.

[500] Health Benefits of Fennel, Rosemary Volatile Oils and their Nano-Forms in Dyslipidemic Rat Model. *Pak J Biol Sci.* 2018 Jan;21(7):348-358.

[501] Timoumi R, Salem IB, Amara I, et al. Protective effects of fennel essential oil against oxidative stress and genotoxicity induced by the insecticide triflumuron in human colon carcinoma cells. *Environ Sci Pollut Res Int.* 2020 Mar;27(8):7957-7966.

[502] Boukhris M, Bouaziz M, Feki I, et al. Hypoglycemic and antioxidant effects of leaf essential oil of Pelargonium graveolens L'Her, in alloxan induced diabetic rats. *Lipids Health Dis.* 2012 Jun 26;11:81.

[503] Zhu LY, Gao YS, Song LZ, et al. [Research on improving memory impairment of blue lavender volatile oil]. *Zhongguo Zhong Yao Za Zhi.* 2017 Dec;42(24):4819-4826.

[504] Qadeer S, Emad S, Perveen T, et al. Role of ibuprofen and lavender oil to alter the stress induced psychological disorders: A comparative study. *Pak J Pharm Sci.* 2018 Jul;31(4(Supplementary)):1603-1608.

[505] Souri F, Rakhshan K, Erfani S, et al. Natural Lavender Oil (Lavandula Angustifolia) Exerts Cardioprotective Effects Against Myocardial Infarction by Targeting Inflammation and Oxidative Stress. *Inflammopharmacology.* 2019 Aug;27(4):799-807.

[506] Xie Q, Wang Y, Zou GL. Protective effects of lavender oil on sepsis-induced acute lung injury via regulation of the NF-κB pathway. *Pharm Biol.* 2022 Dec;60(1):968-978.

Index

www.ingramcontent.com/pod-product-compliance
Lightning Source LLC
Chambersburg PA
CBHW081359270326
41930CB00015B/3353